MARINE MAMMAL POPULATIONS AND OCEAN NOISE

DETERMINING WHEN NOISE CAUSES BIOLOGICALLY SIGNIFICANT EFFECTS

Committee on Characterizing Biologically Significant
Marine Mammal Behavior

Ocean Studies Board

Division on Earth and Life Studies

NATIONAL RESEARCH COUNCIL
OF THE

THE NATIONAL ACADEMIES PRESS
Washington, DC
www.nap.edu

NATIONAL ACADEMIES PRESS 500 Fifth Street, N.W. Washington, DC 20001

NOTICE: The project that is the subject of this report was approved by the Governing Board of the National Research Council, whose members are drawn from the councils of the National Academy of Sciences, the National Academy of Engineering, and the Institute of Medicine. The members of the committee responsible for the report were chosen for their special competences and with regard for appropriate balance.

This study was supported by Grant No. N00014-03-1-0886 between the National Academy of Sciences and the National Oceanographic Partnership Program with support from the National Oceanic and Atmospheric Administration, Office of Naval Research, National Science Foundation, and the Minerals Management Service. Any opinions, findings, conclusions, or recommendations expressed in this publication are those of the author(s) and do not necessarily reflect the views of the organizations or agencies that provided support for the project.

International Standard Book Number: 0-309-09449-6 (Book)
International Standard Book Number: 0-309-54667-2 (PDF)

Additional copies of this report are available from the National Academies Press, 500 Fifth Street, NW, Lockbox 285, Washington, DC 20055; (800) 624-6242 or (202) 334-3313 (in the Washington metropolitan area); Internet, http://www.nap.edu.

Copyright 2005 by the National Academy of Sciences. All rights reserved.

Printed and bound in Great Britain by
Marston Book Services Limited, Oxfordshire

THE NATIONAL ACADEMIES
Advisers to the Nation on Science, Engineering, and Medicine

The **National Academy of Sciences** is a private, nonprofit, self-perpetuating society of distinguished scholars engaged in scientific and engineering research, dedicated to the furtherance of science and technology and to their use for the general welfare. Upon the authority of the charter granted to it by the Congress in 1863, the Academy has a mandate that requires it to advise the federal government on scientific and technical matters. Dr. Bruce M. Alberts is president of the National Academy of Sciences.

The **National Academy of Engineering** was established in 1964, under the charter of the National Academy of Sciences, as a parallel organization of outstanding engineers. It is autonomous in its administration and in the selection of its members, sharing with the National Academy of Sciences the responsibility for advising the federal government. The National Academy of Engineering also sponsors engineering programs aimed at meeting national needs, encourages education and research, and recognizes the superior achievements of engineers. Dr. Wm. A. Wulf is president of the National Academy of Engineering.

The **Institute of Medicine** was established in 1970 by the National Academy of Sciences to secure the services of eminent members of appropriate professions in the examination of policy matters pertaining to the health of the public. The Institute acts under the responsibility given to the National Academy of Sciences by its congressional charter to be an adviser to the federal government and, upon its own initiative, to identify issues of medical care, research, and education. Dr. Harvey V. Fineberg is president of the Institute of Medicine.

The **National Research Council** was organized by the National Academy of Sciences in 1916 to associate the broad community of science and technology with the Academy's purposes of furthering knowledge and advising the federal government. Functioning in accordance with general policies determined by the Academy, the Council has become the principal operating agency of both the National Academy of Sciences and the National Academy of Engineering in providing services to the government, the public, and the scientific and engineering communities. The Council is administered jointly by both Academies and the Institute of Medicine. Dr. Bruce M. Alberts and Dr. Wm. A. Wulf are chair and vice chair, respectively, of the National Research Council.

www.national-academies.org

COMMITTEE ON CHARACTERIZING BIOLOGICALLY SIGNIFICANT MARINE MAMMAL BEHAVIOR

DOUGLAS WARTZOK (*Chair*), Florida International University, Miami
JEANNE ALTMANN, Princeton University, Princeton, New Jersey
WHITLOW AU, University of Hawaii, Manoa
KATHERINE RALLS, Smithsonian Institution, Washington, DC
ANTHONY STARFIELD, University of Minnesota, St. Paul
PETER L. TYACK, Woods Hole Oceanographic Institution, Woods Hole, Massachusetts

Staff

JOANNE BINTZ, Study Director
JENNIFER MERRILL, Study Director
REBECCA NADEL, Christine Mirzayan Science and Technology Policy Intern
DENISE GREENE, Senior Program Assistant
SARAH CAPOTE, Senior Program Assistant
TERESIA WILMORE, Program Assistant
NORMAN GROSSBLATT, Senior Editor

OCEAN STUDIES BOARD

NANCY RABALAIS (Chair), Louisiana Universities Marine Consortium, Chauvin
LEE G. ANDERSON, University of Delaware, Newark
WHITLOW AU, University of Hawaii, Manoa
ARTHUR BAGGEROER, Massachusetts Institute of Technology, Cambridge
RICHARD B. DERISO, Inter-American Tropical Tuna Commission, La Jolla, California
ROBERT B. DITTON, Texas A&M University, College Station
EARL DOYLE, Shell Oil (retired), Sugar Land, Texas
ROBERT DUCE, Texas A&M University, College Station
PAUL G. GAFFNEY, II, Monmouth University, Long Branch, New Jersey
WAYNE R. GEYER, Woods Hole Oceanographic Institution, Woods Hole, Massachusetts
STANLEY R. HART, Woods Hole Oceanographic Institution, Woods Hole, Massachusetts
RALPH S. LEWIS, Connecticut Geological Survey (retired), Hartford
WILLIAM F. MARCUSON III, US Army Corp of Engineers (retired), Vicksburg, Mississippi
JULIAN MCCREARY JR, University of Hawaii, Honolulu
JACQUELINE MICHEL, Research Planning, Inc., Columbia, South Carolina
JOAN OLTMAN-SHAY, Northwest Research Associates, Inc., Bellevue, Washington
ROBERT T. PAINE, University of Washington, Seattle
SHIRLEY A. POMPONI, Harbor Branch Oceanographic Institution, Fort Pierce, Florida
FRED N. SPIESS, Scripps Institution of Oceanography, La Jolla, California
DANIEL SUMAN, Rosenstiel School of Marine and Atmospheric Science, University of Miami, Florida

Staff

SUSAN ROBERTS, Director
JENNIFER MERRILL, Senior Program Officer
DAN WALKER, Senior Program Officer
ALAN B. SIELEN, Visiting Scholar
ANDREAS SOHRE, Financial Associate
SHIREL SMITH, Administrative Coordinator
JODI BACHIM, Research Associate
NANCY CAPUTO, Research Associate
SARAH CAPOTE, Senior Program Assistant

Preface

Biologically significant is an easy modifier to insert into many descriptors, from habitat designations to pharmacological reactions. It has the attributes of a perfectly reasonable modifier. After all, who would object to putting a limit on the great panoply of varied habitats or potential responses encountered in nature? However, when one attempts to distinguish between biologically significant and biologically not significant, the first question is, To whom? The initial choice of range—from habitat to pharmacology—implies the breadth with which this modifier has been used. Biologically significant changes at the habitat level imply alterations in the composition of species that use a habitat. Biologically significant changes at the pharmacological level imply organism changes. Intermediate between those levels are the population (or stock in marine mammal management terms) and the species.

The most basic goal of the Marine Mammal Protection Act (MMPA) (16 U.S.C. 1361) is to maintain marine mammals as a "significant functioning element in the ecosystem of which they are a part." The MMPA translates that ecosystem goal to the population level by aiming to ensure that marine mammal stocks do not fall below or are restored to their optimal sustainable population sizes. Although the main goals of the MMPA are defined at the ecosystem and population levels, its primary focus of regulation is at the level of the individual. When the MMPA was enacted, marine mammal populations were threatened by hunting and by deaths resulting from becoming entangled in nets or otherwise killed in fisheries. The primary regulatory mechanism in the MMPA was a prohibition of the taking of marine mammals; where "take" means to harass, hunt, capture, or kill or attempt to harass, hunt, capture, or kill any marine mammal. The prohibition of taking has reduced the death and injury of marine mammals enough that today many important threats involve habitat degradation and the cumulative effects of harassment. Although harassment is included as a prohibited taking in the MMPA, this prohibition has proved ill suited for protecting marine mammal habitat and regulating cumulative effects.

One approach for protecting marine mammals might be to monitor their populations and initiate protective measures for populations in decline. However, we cannot estimate trends precisely for most marine mammal populations, and by the time a decline is detected, it may be too late. In addition, we also need methods to determine which human activities or natural phenomena are causing population declines or inhibiting population recovery. Many effects of human activities on individual marine mammals occur on a time scale of seconds to years, effects on populations on a scale of years to generations, and effects on ecosystems on a scale of generations to centuries. This report focuses on changes at the population level, but what can be observed are the much faster changes in the behavior and physiology of individuals. The basic goal of this report is to explore the scientific challenge of using short-term observations at the level of individuals to predict effects on populations. Such a predictive model would serve two functions: identifying when the cumulative sum of human effects poses a risk to a population and identifying the activities that pose the greatest risk.

What little we know about behavioral responses of marine mammals to anthropogenic noise highlights the importance of context, including the demographic status of the animals receiving the sound; the characteristics, location, and movement of the sound source; and the location of the animals. The history of the animals is also important: prior exposure to the sound could have resulted in habituation or sensitization. Context includes population status and ecosystem changes; responses that would be insignificant in a population near its carrying capacity can become significant in populations that are depleted or that are encountering multiple stressors, such as El Niño.

Our glimpses into the lives of marine mammals are so short that it is difficult to determine whether the small part of a behavioral reaction we usually can observe is biologically significant. In contrast with Supreme

Court Justice Potter Stewart's statement with respect to pornography, "I know it when I see it" (Jacobellis v. Ohio, 378 U.S. 184, 197 [1964]), the problem in determining the biological significance of marine mammal responses is that often we do *not* know them when we see them. Marine mammals are so hard to observe that we may never see serious problems without studies that are targeted to understand their normal behavior and physiology in the wild. A basic tenet of responsible management and conservation is the need to balance the risks posed by overregulation and those posed by underregulation; the latter carry more weight in conditions of greater uncertainty. The depth of our uncertainty in these issues can make it difficult to calibrate the proper extent of precaution.

A reader who expects this volume to provide a "Eureka" moment of insight into the biological significance of marine mammal responses to noise will be disappointed. That should not come as a surprise. Biological significance has not been well defined in many animal groups that are much more amenable to observation than marine mammals and on which much more data are available. The last few decades have seen a rapid increase in studies of the responses of marine mammals to noise, and there is growing evidence that some sounds play a role in lethal strandings of deep-diving beaked whales, but there is not one case in which data can be integrated into models to demonstrate that noise is causing adverse affects on a marine mammal population. In the case of strandings, the primary data gaps are in our ignorance of the population size and status of beaked whales, and our uncertainty about the number of animals killed or injured. For most other noise effects, the primary source of uncertainty stems from our difficulty in determining the effects of behavioral or physiological changes on an individual animal's ability to survive, grow, and reproduce.

This report contains a conceptual model designed to serve as a roadmap for developing a predictive model that will relate behavioral responses caused by anthropogenic sound to biologically significant, population-level consequences. It identifies the extent of current knowledge and data gaps in each component of the proposed conceptual model to show where research is most needed. In addition to pointing toward a decade-long research agenda for the predictive model, the report suggests management alternatives for the short term and the intermediate term. It also recommends changes in the regulatory structure to include effects of sound on marine mammals within the broader management structure now used exclusively for fisheries. The goal is a common metric for the impact of all human activities on marine mammals and consistent regulation of that impact.

Although a model for predicting the biological significance of different effects cannot be created today, this report offers an approach that can be implemented now to identify, within specified limits, when the responses of marine mammals to anthropogenic noise do not rise to the level of biological significance. The first step in dealing with an apparently intractable problem is to bound it, and this report describes a method for doing that.

Acknowledgments

This report was greatly enhanced by the participants in the workshop held as part of the study. The committee would first like to acknowledge the efforts of those who gave presentations at meetings and thereby helped to set the stage for fruitful discussions in the closed sessions that followed:

JAY BARLOW, Scripps Institution of Oceanography
MELBOURNE BRISCOE, Office of Naval Research
JEAN COCHRANE, US Fish and Wildlife Service
DANIEL P. COSTA, University of California, Santa Cruz
ROGER GENTRY, National Oceanic and Atmospheric Administration National Marine Fisheries Service
WAYNE GETZ, University of California, Berkeley
ROBERT GISINER, Office of Naval Research
DANIEL GOODMAN, Montana State University
BRUCE KENDALL, University of California, Santa Barbara
JAMES KENDALL, Minerals Management Service
S.A.L.M. KOOIJMAN, Vrije Universiteit, Amsterdam
BOB KULL, Parsons
BILL MORRIS, Duke University
TIM RAGEN, Marine Mammal Commission
L. MICHAEL ROMERO, Tufts University
GORDON SWARTZMAN, Applied Physics Laboratory

SHRIPAD TULJAPURKAR, Stanford University
JAMES YODER, National Science Foundation

This report has been reviewed in draft form by persons chosen for their diverse perspectives and technical expertise, in accordance with procedures approved by the National Research Council's Report Review Committee. The purpose of this independent review is to provide candid and critical comments that will assist the institution in making its published report as sound as possible and to ensure that the report meets institutional standards of objectivity, evidence, and responsiveness to the study charge. The review comments and draft manuscript remain confidential to protect the integrity of the deliberative process. We wish to thank the following for their participation in their review of this report:

HAL CASWELL, Woods Hole Oceanographic Institution, Woods Hole, Massachusetts
DAN COSTA, University of California, Santa Cruz
COLLEEN REICHMUTH KASTAK, University of California, Santa Cruz
ROBERT KNOX, University of California, San Diego
ROBERT KULL, Parsons, Norfolk, Virginia
PAUL NACHTIGALL, University of Hawaii, Kailua
DON SINIFF, University of Minnesota, St. Paul
NINA YOUNG, The Ocean Conservancy, Washington, DC

Although the reviewers listed above have provided many constructive comments and suggestions, they were not asked to endorse the conclusions or recommendations, nor did they see the final draft of the report before its release. The review of this report was overseen by **John Dowling**, Harvard University, and **Andrew Solow**, Woods Hole Oceanographic Institution, appointed by the National Research Council, who were responsible for making certain that an independent examination of the report was carried out in accordance with institutional procedures and that all review comments were carefully considered. Responsibility for the final content of the report rests entirely with the committee and the institution.

Contents

EXECUTIVE SUMMARY   1

1   INTRODUCTION   13
    Defining the Problem, 13
    History of National Research Council Reports, 15
    Call for A New National Research Council Study, 18

2   CURRENT STATE OF KNOWLEDGE OF BEHAVIORAL
    AND PHYSIOLOGICAL EFFECTS OF NOISE ON
    MARINE MAMMALS   23
    Behavioral Responses to Acoustic Stimuli, 23
    Physiological Responses to Acoustic Stimuli, 29
        Auditory Damage, 29
        Nonauditory Effects of Sound, 31
            Resonance Effects, 31
            Rectified Diffusion, 32
    Progress on Earlier National Research Council Recommendations, 34

3   HOW TO GET FROM ACOUSTIC DISTURBANCE TO
    POPULATION EFFECTS   35
    Current Data Collection Efforts, 46
    Data Needed to Determine Physiological Responses to Acoustic
        Stimuli, 47

Physiological Stress Effects, 48
Toxicology, 56
Data Needed to Describe Marine Mammal Populations, 56
Individuals to Populations: Using Models to Improve Understanding, 57
   Uses of Models: Prediction and Exploration, 58
   Demographic Models, 61
   Individual-Based Models, 62
   Categorical or Qualitative Models, 64
   Expert Opinion, 65
   Risk Assessment, 65

4   RATIONAL MANAGEMENT WITH INCOMPLETE DATA   69
Potential Biological Removal, 71
Extension of PBR, 73
Determination of Nonsignificant Impact, 79

REFERENCES   87

APPENDIXES

A   Committee and Staff Biographies   101
B   Acronyms   105
C   Workshop Agenda and Participants List   107
D   Draft Conceptual Plan for Workshop Discussion   113
E   Scientific and Common Names   125

Executive Summary

The transition from wind-driven to mechanized shipping became the first step in what was to be a continued increase in the introduction of sound into the oceans. The oceans are much less transparent to light than to sound; as a result, many marine species use sound rather than light to navigate and communicate. Over the last 40 million years, marine mammals have evolved specializations for using underwater sound. The initial introduction of the propulsion sound of ships was unintentional, but engineers and scientists have also learned, with the development of sonar, how to use sound intentionally for underwater communication, navigation, and research. At some point as humans introduce more sound into the oceans, the conflict with evolutionarily-adapted marine mammal sound-sensing systems seems inevitable. Attention has been drawn to this issue through a series of marine mammal strandings, lawsuits, legislative hearings, and National Research Council (NRC) reports (1993, 2000, and 2003b) and, most recently, the draft report of the US Commission on Ocean Policy (2004).

Two earlier National Research Council reports (1994, 2000), while addressing biological issues of marine mammals and noise, also made recommendations that affected federal legislation and its implementation. The first was issued in 1994 in response to the feasibility test of a proposal to track global warming by monitoring the speed of an acoustic signal across an ocean basin (Munk et al., 1994). The feasibility test was to have set the stage for the full Acoustic Thermometry of the Ocean Climate (ATOC) experiment, but because of concerns over possible effects on marine mammals only a limited deployment of ATOC was attempted. The 1994 report recommended that there be legislative distinction between different types of "taking" and that the regulatory agencies streamline the permitting process for activities that did not kill or capture marine mammals. Additional streamlining was recommended for nonlethal activities that have negligible effects. The 2000 National Research Council report reviewed the marine mammal research program that was a component of the limited ATOC deployment. In *Marine Mammals and Low-frequency Sound: Progress Since 1994*, the committee noted that the 1994 amendments to the Marine Mammal Protection Act (MMPA) addressed some of the issues raised in the 1994 NRC report. The 1994 amendments introduced two levels of takes by harassment under the MMPA—level A and level B harassment. Level A harassment was defined in the 1994 amendments as "any act of pursuit, torment, or annoyance which has the potential to injure a marine mammal or marine mammal stock in the wild." Level B harassment was defined as "any act of pursuit, torment, or annoyance which has the potential to disturb a marine mammal or marine mammal stock in the wild by causing disruption of behavioral patterns including, but not limited to, migration, breathing, nursing, breeding, feeding, or sheltering." However, the 2000 National Research Council report emphasized the importance of a criterion for significance of disruption of behavior (pg. 68):

> It does not make sense to regulate minor changes in behavior having no adverse impact; rather, regulations must focus on significant disruption of behaviors critical to survival and reproduction.

The report (pg. 69) recommended redefining level B harassment as any act that

> has the potential to disturb a marine mammal or marine mammal stock in the wild by causing meaningful disruption of biologically significant activities, including but not limited to, migration, breeding, care of young, predator avoidance or defense, and feeding.

Since the report was issued, the term *biologically significant* has been used in discussions of the 2003-2004 reauthorization of the MMPA (House Report 108-464). The US National Marine Fisheries Service (now National Oceanic and Atmospheric Administration [NOAA] Fisheries) has also used the term in decisions to grant incidental harassment authorizations. Scientific investigation and description of what would constitute "biologically significant" have not been pursued in a comprehensive manner.

The charge to the present National Research Council committee (Box ES-1) was to clarify the term *biologically significant*. In the broadest sense, it is a straightforward charge. An action or activity becomes biologically significant to an individual animal when it affects the ability of the animal to grow, survive, and reproduce. Those are the effects on individuals that can have population-level consequences and affect the viability of the species. However, those effects are separated in time and usually in space from the precipitating event. What can be observed, with difficulty in the case of marine mammals, are the direct behavioral and in some cases physiological responses of individual animals.

It was recognized that the definition of level B harassment proposed in the 2000 report introduced two kinds of biological significance: one with respect to animal activities, stated directly, and the other implied in the "meaningful disturbance" of those activities. On reflection, it became clear that wild animals rarely engage in activities that are not biologically significant (even play is not frivolous [Bekoff and Byers, 1998]), so the primary

---

BOX ES-1
Statement of Task

In its 2000 report, *Marine Mammals and Low-frequency Sound*, the National Research Council recommended that the Marine Mammal Protection Act definition of "Level B harassment should be limited to meaningful disruption of biologically significant activities that could affect demographically important variables such as reproduction and longevity." Recognizing that the term "biologically significant" is increasingly used in resource management and conservation plans, this study will further describe the scientific basis of the term in the context of marine mammal conservation and management related to ocean noise. Based on input from a scientific workshop, consideration of the relevant literature, and other sources, the committee will produce a brief report that reviews and characterizes the current scientific understanding of when animal behavior modifications induced by transient and non-transient ocean acoustic sources, individually or cumulatively, affect individuals in ways that have negative consequences for populations.

concern should be with determining when human activity elicits behavioral or physiological responses in marine mammals that rise to the level of biological significance.

Changes in behavior that lead to alterations in foraging efficiency, habitat abandonment, declines in reproduction, increases in infant mortality, and so on are difficult to demonstrate in terrestrial animals, including humans, and are much more difficult to demonstrate in animals that may only rarely be observed in their natural environment.

A CONCEPTUAL MODEL TO ADDRESS POPULATION CONSEQUENCES OF ACOUSTIC DISTURBANCE

A conceptual model is proposed that identifies the different stages required to move from marine mammal behavior to a determination of population effects of behavioral change. The model first characterizes an acoustic signal, the resulting behavioral change, and a determination of the "life function" or activity affected. It then describes the resulting change in vital rate, such as life span, and finally suggests population effects—effects on following generations. "Transfer functions" connect the variables. A transfer function is essentially a relationship that allows one to estimate, for example, how a change in migration route leads to a reduction in reproductive success. It was quickly recognized that the high-priority research identified in the earlier National Research Council reports (1994, 2000, 2003b) is essential for building the first stages of the model.

RECOMMENDATION 1: The high-priority research identified by the National Research Council (1994, 2000, 2003b) should be completed. That research is essential for the model proposed in this report.

Through discussions before and during the public workshop held at the National Academies in March 2004, a consensus was reached that the proposed conceptual model includes the components needed to develop a predictive model to determine the biological significance of behavioral change. However, there was also a consensus that we are a decade or more away from having the data and understanding of the transfer functions needed to turn such a conceptual model into a functional, implementable tool.

RECOMMENDATION 2: A conceptual model, such as that described in this report, should be developed more fully to help to assess impacts of acoustic disturbance on marine mammal populations. Development of such a model will allow sensitivity analysis that can be used to focus, stimulate, and direct research on appropriate transfer functions.

To enhance such a model and progress toward determining population effects of acoustic disturbance, all available sources of data on marine mammal behavior and reactions to noise will need to be accessed. In addition to results of normal scientific studies, a veritable wealth of data on marine mammals is collected in compliance with federal regulatory requirements, but those data are not being accessed or used beyond the original intent of their collection (such as for permit issuance). A data-coordination effort could provide substantial benefits and improve our knowledge of marine mammal distribution, critical habitats, behaviors, population estimates, and other items essential for the modeling effort. Although data coordination would be difficult to implement, over the long term the value added by improving the organization and accessibility of data collected for these purposes would provide an efficient means of extracting invaluable information, at relatively small additional cost, for improving understanding and management. Such leveraging of diverse data collection efforts would represent an efficient use of resources and public funding. For example, the UK Joint Natural Conservation Committee has summarized sighting data from commercial seismic surveys, which help in evaluating avoidance responses (Stone, 2001, 2003).

RECOMMENDATION 3: To assist in the development of the conceptual model, a centralized database of marine mammal sightings and their responses to anthropogenic sound in the ocean should be developed and should include

- **Published peer-reviewed papers in the scientific literature.**
- **Government technical reports.**
- **Data submitted to NOAA Fisheries and the US Fish and Wildlife Service in permit applications.**
- **Data submitted by industry to the Minerals Management Service for regulating off-shore hydrocarbon exploration and production.**

- **All relevant data accumulated by all federal agencies in the course of their research and operational activities, including monitoring.**

To facilitate the integration of data from the various sources, federal agencies need to develop standardized data-reporting formats. Survey data should include locations where marine mammals were detected and the track lines when personnel were monitoring for marine mammals, regardless of whether any were sighted. All data entered into such an integrated database must be coded as to quality, and peer-reviewed data and interpretations should be rated highest.

The biological significance of the behavioral response of an animal to an acoustic stimulus is modulated by many seasonal and environmental factors. For example, the lengthening of a foraging trip from a rookery that would be of no particular significance during a normal year could rise to the level of biological significance during an El Niño year. Allostasis, the maintenance of an animal's physiological stability in spite of change, is a useful way to conceptualize the integration of short-term and cumulative stress and thereby to determine the possible additional effects of anthropogenic noise on marine mammals. Although data for marine mammals are lacking, serum hormone concentrations have been shown to be good measures of stress in terrestrial animals. For animals in which blood sampling is impractical, fecal sampling has been used successfully and is now being applied to some marine mammals. Preliminary studies measuring glucocorticoids in hair samples and enhanced synthesis of RNA coding for stress-induced proteins in skin samples merit further development. Measures of stress may provide critical information on marine mammal physiological status and change in response to disturbance by acoustic and other stimuli.

Correlational observations of behavioral responses to noise and other stressors have indicated general trends in such responses and in some cases have highlighted subjects of concern. To calibrate an animal's response to a stimulus as required for a predictive model, correlational observations must be replaced with controlled-exposure, dose-response experiments. Such an approach allows researchers to go beyond observational study and determine statistically the likelihood of a particular reaction to a given acoustic stimulus. In marine mammals, such experiments are only beginning to be conducted. There is a potential for collecting both behavioral and physiological data during controlled-exposure experiments. The resulting data will be essential for integrating behavior and physiology in models of the population consequences of acoustic disturbance.

Additional development of data-logging technology is necessary for support of controlled-exposure experiments. Data-logging packages should be modified to incorporate blood sampling during controlled-exposure experiments. Initial studies on Weddell seals (*Leptonychotes weddellii*) would be particularly useful in as much as research on their blood chemistry during free dives has already been conducted (Hill, 1986). Eventually the packages would benefit from new less-invasive methods for collecting blood and conducting "on-board" blood-chemistry analysis to record responses of animals in situations less unusual than that of the Weddell seal—a situation in which the animal can be handled before and after tagging.

> **RECOMMENDATION 4: The use of glucocorticoid and other serum hormone concentrations to assess stress should be developed, validated, and calibrated for various marine mammal species and age-sex classes and conditions. Dose-Response curves for those indicators as a function of sound characteristics need to be established. Development of a sampling package that could take blood samples on a controlled basis and stabilize hormones for later analysis or process samples "on-board" for corticosteroids at various stages of a CEE would be invaluable for determining the stress that the sound is producing. The use of fecal sampling to measure condition or stress needs to be investigated further and developed. Research efforts should seek to determine whether reliable long-term stress indicators exist and, if so, whether they can be used to differentiate between noise-induced stress and other sources of stress in representative marine mammal species (this recommendation was also made in NRC, 2003b).**

Although the full predictive model of the path from acoustic stimulus to population effect is unattainable in the near term, various modeling techniques can enhance our understanding of the components of the larger model. One approach involves demographic models in which age- or stage-specific developmental, behavioral, or physiological characteristics of individuals are used to explore changes in population dynamics (Caswell,

2001). Another approach involves individual-based models that can be used to infer population responses by tracing the life history of individuals. For a number of nonmarine mammal species, individual-based models include physiology and behavior; such models have provided insight into how ecological change and human disturbance have altered demographic variables. Although a thorough, detailed model is not now possible for any marine mammal species, this approach can be used to provide preliminary understanding and to identify the most crucial gaps in available data.

Qualitative or categorical modeling that characterizes the strength of links between stimulus and response, response and function, and function and demography on a simple low-medium-high basis can be useful. A focused effort is needed on a modeling exercise that should include quantitative demographic models, individual-based models, and qualitative categorical models. Such an effort should start with, and be calibrated against, expert opinion. The effort should

- Probe how successfully current knowledge could be applied.
- Identify crucial gaps in our knowledge.
- Encourage and provide structure for interdisciplinary synthesis.
- Require that all modeling efforts be explicit about uncertainty and the consequences of uncertainty.
- Require that all models clearly state their limited purpose and evaluate both their strengths and their shortcomings.
- Assess the risk for the species being modeled if the model is to be used for management decisions.

Exploratory models could help to bridge the gap between changes in the physiology and behavior of individuals in response to sound and demographic effects at the population level. Demographic models might be used in an exploratory way to help to bound the problem and establish thresholds for different species. Individual-based models may provide a method for exploring the consequences of changes in individual behavior and social interactions. Those modeling approaches could be used, individually or in combination, to provide greater understanding of the problem, look for important thresholds, speculate on the likely outcome of hypothesized changes, and develop a conceptual framework for formulating management guidelines.

RECOMMENDATION 5: Several marine mammal species for which there are good long-term demographic and behavioral data on individuals should be selected as targets of an intensive exploratory modeling effort that would develop a series of individual-based models and stage- or age-structured demographic models for the species as appropriate. NOAA Fisheries should bring together an independent, interdisciplinary panel of modelers and relevant empirical scientists that would meet periodically to pursue the modeling effort collaboratively in an iterative and adaptive manner with the long-term goal of developing tools to support informed, practical decision-making.

As noted, the full predictive model is at least a decade away from coming to fruition, and the management requirements involved in addressing concerns over ocean-noise effects on marine mammals are extremely pressing. Efforts are under way to address the long-term goal of producing the predictive model outlined here, but an interim plan is needed. One strategy is to implement a management regimen that uses available data, agreed upon management goals, and a conservative approach to the insufficiencies of the available data. The regimen should encourage data acquisition to reduce uncertainty. The NOAA Fisheries Potential Biological Removal (PBR) model is such an example.

RECOMMENDATION 6: A practical process should be developed to help in assessing the likelihood that specific acoustic sources will have adverse effects on a marine mammal population by disrupting normal behavioral patterns. Such a process should have characteristics similar to the Potential Biological Removal model, including

- Accuracy,
- Encouragement of precautionary management—that is more conservative (smaller removal allowed)—when there is greater uncertainty in the potential population effects of induced behavioral changes,
- Being readily understandable and defensible to the public, legal staff, and Congress,

- An iterative process that will improve risk estimates as data improve,
- Ability to evaluate cumulative impacts of multiple low-level effects, and
- Construction from a small number of parameters that are easy to estimate.

The PBR model has the potential for being applied more widely than it is now. So far, for most species it has incorporated only direct fishery mortalities and serious injuries in the determination of biological removal. Indirect fishery mortalities, nonfishery mortalities, and mortality equivalents for injury and disruption need to be added to the biological removal in the model to encompass the multitude of effects, including acoustic effects, of human activities on marine mammal populations.

RECOMMENDATION 7: Improvements to PBR are needed to reflect total mortality losses and other cumulative impacts more accurately:

- NOAA Fisheries should devise a revised PBR in which all sources of mortality and serious injury can be authorized, monitored, regulated, and reported in much the same manner as is currently done by commercial fisheries under Section 118 of the MMPA.
- NOAA Fisheries should expand the PBR model to include injury and behavioral disturbance with appropriate weighting factors for severity of injury or significance of behavioral response (cf. NRC, 1994, pg. 35).

Current knowledge is insufficient to predict which behavioral responses to anthropogenic sounds will result in significant population consequences for marine mammals. The predictive model and even the proposed revisions to PBR will take years to implement. In the interim, those who introduce sound into the marine environment and those who have responsibility for regulating sound sources need a system whereby reasonable criteria can be set to determine whether a particular sound source will have a non-significant effect on marine mammal populations. Collectively, there is sufficient expert knowledge and there are extensive databases to establish such a system and to set the criteria conservatively enough for there to be broad agreement on the nonsignificant effect criterion. An example of a preliminary application of the approach is the impact-likelihood risk-evaluation matrices developed for typical acoustic equipment used on research vessels in the Antarctic (SCAR, 2004).

RECOMMENDATION 8: An intelligent-decision system should be developed to determine a de minimis standard for allowing proposed sound-related activities. An expert-opinion panel should be constituted to populate the proposed system with as many decision points as current information and expert opinion allow. The system should be systematically reviewed and updated regularly.

The goal of this report is to provide a method for clarification of the concept of biologically significant disturbance. The recommendations made here are intended to provide both a long-term, well-supported, and valid solution and a near-term problem-solving strategy to assist resource managers in coping with this difficult and complex issue.

1

Introduction

DEFINING THE PROBLEM

Throughout human history, the oceans have been important for transportation and commerce, for their biological and physical resources, and for defense. The vast expanse of the oceans precluded significant human impact until the coming of the industrial revolution when the transition from wind-driven to mechanized shipping became the first step in what was to be a continued increase in the unintentional and then—with the development of sonar—intentional introduction of sound into the oceans. Because of the low loss characteristic of sound transmission compared with light transmission, the use of sound had developed evolutionarily as the predominant long-range sensory modality for marine mammals. As engineers and scientists learned to appreciate the properties of acoustic propagation in the sea, they introduced sound sources to communicate and to detect objects in the oceans or on or below the seafloor. At some point, as humans use the oceans more and increase anthropogenic sound in the oceans, the conflict with evolutionarily adapted marine animals' sound-sensing systems seems inevitable.

Over 90% of global trade uses the sea for transportation. Shipping is the dominant source of sound in the world's oceans in the range from 5 to a few hundred Hertz. At other frequencies, anthropogenic sound does not predominate in the ocean sound-energy budget, but it can have important local effects (NRC, 2003b). Seismic air guns associated with geophysical exploration for locating new oil and gas deposits run hundreds of thousands of miles of survey lines in the Gulf of Mexico alone each year. Commercial sonar systems are on all but the smallest pleasure craft and permit safer boating and shipping and more productive fishing. Military sonar systems are important for national defense. Ocean noise from human and natural sources can also originate in the air, as in sonic booms, lightning, and wind (NRC, 2003b).

The intentional and unintentional introduction of sound in the oceans associated with activities beneficial to humans has known deleterious effects on individual marine mammals. Mass strandings of beaked whales, defined as strandings involving two or more animals other than female-calf pairs (Geraci and Lounsbury, 1993), in some cases have clearly been associated with the use of midrange tactical military sonar (D'Amico, 1998; Evans and England, 2001; Jepson et al., 2003). Beluga whales (*Delphinapterus leucas*) have strong and prolonged behavioral responses to icebreakers 50 km away under some circumstances (LGL and Greeneridge, 1986; Cosens and Dueck, 1988; Finley et al., 1990). Gray whales (*Eschrichtius robustus*) and killer whales (*Orcinus orca*) have shown multiyear abandonment of critical habitats in response to anthropogenic noise (Bryant et al., 1984; Morton and Symonds, 2002). Although there are many documented, clearly discernible responses of marine mammals to anthropogenic sound, responses are typically subtle, consisting of shorter surfacings, shorter dives, fewer blows per surfacing, longer intervals between blows, ceasing or increasing vocalizations, shortening or lengthening vocalizations, and changing frequency or intensity of vocalizations. Some of those changes become statistically significant for a given exposure, such as increases in descent rate and increases or decreases in ascent rate of northern elephant seals (*Mirounga angustriostris*) in response to Acoustic Thermometry of the Ocean Climate (ATOC) signals (Costa et al., 2003). But it remains unknown when and how these changes translate into biologically significant effects—effects that have repercussions for the animal beyond the time of disturbance, effects on the animal's ability to engage in essential activities, and effects that have potential consequences at the population level.

The basic goal of marine mammal conservation is to prevent human activities from harming marine mammal populations. The threat from commercial whaling was obvious, but it is more difficult to estimate the population consequences of activities that have less immediately dramatic outcomes, such as those with indirect or small but persistent effects. The life histories and habitat of marine mammals compound the difficulties.

INTRODUCTION

Marine mammals are long-lived and slow to mature. The young of many species are dependent for long periods. They are highly social, have behavioral plasticity, and have complex processes of behavioral development. Many of their behaviors occur underwater, where they are difficult to document, and that makes it particularly hard to estimate the effects of a short-term exposure as they ripple through the lifetime of an individual or as the effects on different individuals ripple through the population. Even extreme effects, including death, are not necessarily observed.

With the exception of the beaked whale strandings, connections between anthropogenic sound in the oceans and marine mammal deaths have not been documented. In the presence of clear evidence of lethal interactions between humans and marine mammals in association with fishing and vessel collisions (Clapham et al., 1999; Laist et al., 2001), the absence of such documentation has raised the question of the relative importance of sound in the spectrum of anthropogenic effects on marine mammal populations. Anthropogenic ocean noise is thought not to be a factor in any of the recent major declines in marine mammal populations, such as Steller sea lions (*Eumetopias jubatus*; NRC, 2003a), harbor seals (*Phoca vitulina*; Pitcher, 1990), fur seals (York, 1987), and Aleutian Island sea otters (*Enhydra lutris*; Doroff et al., 2003). No scientific studies have conclusively demonstrated a link between exposure to sound and adverse effects on a marine mammal population. These considerations have led to alternative assessments of the effects of sound on marine mammals. On the one hand, sound may represent only a second-order effect on the conservation of marine mammal populations; on the other hand, what we have observed so far may be only the first early warnings or "tip of the iceberg" with respect to sound and marine mammals.

HISTORY OF NATIONAL RESEARCH COUNCIL REPORTS

The National Research Council has produced three reports on the effects of noise on marine mammals, in 1994, 2000, and 2003. The primary goal of the first, *Low-Frequency Sound and Marine Mammals: Current Knowledge and Research Needs*, was to address the specific issues raised by the Heard Island Feasibility Test, which sought to "establish the limits of usable, long-range acoustic transmissions" (Munk et al., 1994). The feasibility test was preliminary to the ATOC experiment. The ATOC project proposed to measure the speed of sound across ocean basins as a way to monitor global climate change, and it required long-range transmissions of underwater sound regularly from several sites for decades. The 1994 report recommended research with respect to low-frequency (1- to 1,000-Hz) sound and marine mammals that was needed before a full deployment of ATOC. The report also noted that regulation of marine mammal research impeded exactly the type of research needed to determine if anthropogenic noise is detrimental to the animals. The report included an entire chapter on regulatory issues (NRC, 1994).

The Marine Mammal Protection Act (MMPA, 16 U.S.C. 1361-1401 et seq.) enacted in 1972 is the legal instrument of the US federal government for protection of marine mammals. The 1994 National Research Council report was concerned that the statutory term *harassment*, included in the MMPA but undefined in regulation, was "being interpreted through practice to include any action that results in an observable change in the behavior of a marine mammal" (Swartz and Hofman, 1991). The report pointed out (pg. 28) that

> As researchers develop more sophisticated methods for measuring the behavior and physiology of marine mammals in the field (e.g., via telemetry), it is likely that detectable reactions, however minor and brief, will be documented at lower and lower received levels of human-made sound. . . . In that case, subtle and brief reactions are likely to have no effect on the well-being of marine mammal individuals or populations.

The report recommended that legislative distinctions be made between different types of taking and that the regulatory agencies streamline the permitting process for activities that do not kill or capture marine mammals. Additional streamlining should be considered for nonlethal activities that have negligible effects. Agencies were encouraged to regulate within the context of total human impacts on marine mammals—including fisheries, shipping, the oil and gas industry, and research activities—and to expend their primary effort on activities with the greatest potential for harm.

The 2000 National Research Council report, *Marine Mammals and Low-frequency Sound: Progress Since 1994*, noted that the 1994 amendments to the MMPA addressed some of the issues raised in the 1994 report. The 1994 amendments introduced two levels of disturbance that are considered regulated takings—level A and level B harassment. Level A harassment is "any act of pursuit, torment, or annoyance which has the potential to injure a marine mammal or marine mammal stock in the wild." Level B harassment is "any act of pursuit, torment, or annoyance which has the potential to disturb a marine mammal or marine mammal stock in the wild by causing disruption of behavioral patterns including, but not limited to, migration, breathing, nursing, breeding, feeding, or sheltering." However, the 2000 National Research Council report continued to emphasize the importance of a criterion for significance of disruption of behavior (pg. 68):

> It does not make sense to regulate minor changes in behavior having no adverse impact; rather regulations must focus on significant disruption of behaviors critical to survival and reproduction.

The report recommended a redefinition of level B harassment as any act that (pg. 69)

> has the potential to disturb a marine mammal or marine mammal stock in the wild by causing meaningful disruption of biologically significant activities, including but not limited to, migration, breeding, care of young, predator avoidance or defense, and feeding.

Since the report was issued, the term *biologically significant* has been used in discussions of the 2003-2004 reauthorization of the MMPA (House Report 108-464). The US National Marine Fisheries Service (now National Oceanic and Atmospheric Administration [NOAA] Fisheries) has used the term in decisions to grant incidental harassment authorizations, but scientific investigation and description of what would constitute "biologically significant" have not been pursued in a comprehensive manner.

The 2003 National Research Council report, *Ocean Noise and Marine Mammals*, attempted to quantify the world ocean-noise budget between 1 and 200,000 Hz with particular attention to habitats that are important to marine mammals (NRC, 2003b). The basic question it addressed was the overall impact of human-made sound on the marine environment. The somewhat unsatisfactory answer was that the overall impact is unknown but there is cause for concern. It was noted that total energy contribution is not the best currency to use in determining the potential impact of human-made sound on marine organisms. The report offered a number of recommendations; the overarching one was the need to understand better the characteristics of ocean noise, particularly from human-made noise, and its potential effects on marine life, especially effects that may have population consequences.

Thus, each of the three previous National Research Council reports has recommended research to resolve critical uncertainties about the effects of noise on marine mammals. All three highlighted the need for research in behavioral ecology, auditory physiology and anatomy, nonauditory effects of sound, effects of sound on prey of marine mammals, and development of new techniques for measuring the effects of sound on marine mammals. The 2003 report also recommended research on sources and modeling of ocean noise. Some of the recommendations have led to research that has greatly reduced the data gap. For example, the 1994 and 2000 reports recommended experiments to determine acoustic exposures that would lead to temporary shifts in the threshold of hearing in marine mammals. In the last decade, several laboratories have conducted such experiments (Kastak et al., 1999; Finneran et al., 2000, 2002; Schlundt et al., 2000; Nachtigall et al., 2003, 2004), and there is much less uncertainty in modeling the exposures that start to cause physiological effects on hearing in the seal and small-toothed whale species that have been tested. There has been partial progress on other recommendations. For example, the 1994 report recommended the development of tags to record physiological characteristics, behavior, location, and sound exposure. In the last decade, tags have been developed to record all the features recommended (Burgess, 2001; Johnson and Tyack, 2003) except physiological measures. For many of the other research recommendations research is being conducted, but progress has been slow enough to warrant the establishment of a targeted research program.

The 2000 and 2003 National Research Council reports recommended better coordination between federal regulatory agencies and science-funding agencies to develop a multidisciplinary research program that would judge the quality of proposals with peer review. There has been little progress on those programmatic recommendations, and the present committee re-emphasizes that progress in critical research requires that the federal government develop and fund a dedicated multidisciplinary research program on the subjects in question.

CALL FOR A NEW NATIONAL RESEARCH COUNCIL STUDY

The recommendations of the 2000 National Research Council report have received great attention and been applied by regulators, legislators, and permit applicants to describe level B harassment under the MMPA. The vagaries associated with the term *biologically significant behaviors* and what constitutes "meaningful" disruption of those behaviors have been problematic. In light of the litigious and legislative environment of the issue of the disturbance of marine mammals, several federal agencies (including the Office of Naval Research, the National Science Foundation, the Minerals Management Service, and NOAA), working through the National Oceanographic Partnership Program, requested that the National

Research Council undertake a study to clarify the meaning of the term used in the 2000 report. Which possible effects have population consequences? If we don't know, how can we determine them? The agencies, recognizing that effects will be biologically significant at individual and population levels, requested guidance from the present committee in making those determinations. At the individual level, the biological significance of an effect must be judged by changes in the ability of an animal to grow, survive, and reproduce. The population effect involves the cumulative impact on all individuals affected. The committee's charge, developed with those considerations in mind is shown in Box 1-1.

After discussion of and deliberation on the task statement, the committee recognized that the definition of level B harassment proposed in the 2000 report introduced two kinds of biological significance: one, with respect to animal activities, stated directly, and the other implied in the "meaningful disturbance" of those activities. On reflection, it became clear that animals in the wild rarely spend substantial amounts of time engaging in activities that are not biologically significant. Even seemingly frivolous

---

BOX 1-1
Statement of Task

In its 2000 report, *Marine Mammals and Low-frequency Sound*, the National Research Council recommended that the Marine Mammal Protection Act definition of "Level B harassment should be limited to meaningful disruption of biologically significant activities that could affect demographically important variables such as reproduction and longevity." Recognizing that the term "biologically significant" is increasingly used in resource management and conservation plans, this study will further describe the scientific basis of the term in the context of marine mammal conservation and management related to ocean noise. Based on input from a scientific workshop, consideration of the relevant literature, and other sources, the committee will produce a brief report that reviews and characterizes the current scientific understanding of when animal behavior modifications induced by transient and non-transient ocean acoustic sources individually or cumulatively affect individuals in ways that have negative consequences for populations.

activities, such as play, can be biologically significant (Bekoff and Byers, 1998). Therefore, the primary concern should be with determining when human activity elicits behavioral or physiological responses in marine mammals that rise to the level of biological significance. Population consequences of behavioral change result from the accumulation of responses of individuals. In some cases, thousands of behavioral effects accumulated over years may be necessary for any population consequences; in other cases, a single instance of behavioral response may have the potential for population consequences.

FINDING: As opposed to the definition of biologically significant activities, whose disruption can constitute harassment, the crucial determination is of when behavioral or physiological responses result in deleterious effects on the individual animals and the population.

The statement of task incorporates two issues that had been concerns of earlier National Research Council reports. One is the difference between statistically significant and biologically significant changes in behavior. As more subtle behavioral changes become capable of being observed, it is inevitable that exposure to noise will result in statistically significant changes in one or more of the observed behaviors, but it is not equally certain that the changes will have any biological significance either for the individual or for the population.

The second issue is the linking of short-term behavioral changes to possible consequences at the population level. How does one determine whether an acoustic disturbance can, or does, result in a change in population structure, distribution, or, ultimately, survival? In the absence of any comprehensive model for relating acoustic disturbance to population response with due consideration of all the intermediary steps and processes, the committee developed a conceptual model that, when supplemented with data, would facilitate the recognition of population effects of acoustic disturbance. The model includes an indication of the current state of knowledge and was designed to allow sensitivity analysis that can focus, stimulate, and direct research.

To elaborate the model, identify deficiencies, and summarize current understanding, the committee held a focused public workshop (Appendix C). Workshop panel members were presented with the conceptual model, named the Population Consequences of Acoustic Disturbance (PCAD) model (Appendix D), described in Chapter 2, and asked to apply their expertise in such fields as epidemiology and population biology. Workshop participants discussed the PCAD model—they related it to existing models, identified weaknesses in it, provided an assessment of data available to achieve its objectives, and evaluated the probability of achieving a predictive model in the next decade, given the current understanding of the processes linking behavior to vital rates and given the missing, but required, data. Participants agreed that the model provided a good basis for encompassing the components of the problem, defining needed data, and identifying the research agenda for the next decade. The consensus of the participants, both in their presentations and in breakout groups, was that the model incorporated the necessary components to become a predictive model when sufficient data became available. Workshop discussions of a number of topics improved the information and depth of analysis incorporated in this report, such as the examples of allostasis and the comparison of capital with income pinniped breeders. The initial model did not include the assessments of current knowledge of either the major categories of responses or the transfer functions. Those functions were assigned by the committee after the input of the workshop participants.

This report is the culmination of the workshop presentations, the public dialogues that ensued, and the committee's deliberations. The participants in the workshop made it clear that current empirical data and theoretical knowledge are insufficient to accomplish all the goals of the committee. Therefore, this report offers recommendations intended to provide a roadmap for the development of a predictive model of the effects of ocean noise on marine mammal populations and presents suggestions for temporary measures for regulating the effects until a predictive model is developed and tested.

FINDING: A conceptual model can assist in the understanding of acoustic disturbance of marine mammals and possible effects on populations of them. However, the paucity of data prevents such a model from having a predictive role now.

2

Current State of Knowledge of Behavioral and Physiological Effects of Noise on Marine Mammals

BEHAVIORAL RESPONSES TO ACOUSTIC STIMULI

Various approaches have been used to study behavioral responses of marine mammals to acoustic exposure. Observational studies have been used to correlate distributional or behavioral effects on uncontrolled human activities. That approach is particularly suited to the large spatial or temporal scales over which there may be consistent variation in human activities. For example, Bryant et al. (1984) collated sighting data from surveys of gray whales in one of their breeding lagoons. They reported fewer gray whales sighted after a saltworks started dredging and shipping in the lagoon. Gray whales apparently abandoned the lagoon during this activity, and took several years to start using the lagoon again after the saltworks ceased operating. Although long-term abandonment of critical gray whale breeding habitat clearly reaches the threshold of biological significance, it has not been demonstrated that it impeded the recovery of the population. Morton and Symonds (2002) report a significant decline in sightings of killer whales during a 5-year period when acoustic-harassment devices were operated in an area of water about 10 km × 10 km in an archipelago. The acoustic-harassment devices have a source level of about 194 dB re 1 µPa at 1m and are designed to be loud enough to deter pinnipeds from breaking into fish farms to feed, but they have unintended consequences for inshore cetaceans. The exposures that caused an avoidance reaction in the killer whales are not known—a common problem in correlational studies when precise relationships between acoustic stimuli and behavioral responses are obscure.

Researchers have addressed concerns that marine mammals might avoid intense sounds. Some census studies have towed hydrophones through areas with commercial seismic surveys. Rankin (1999) and Norris, et al. (2000) found no association between the signal-to-noise ratio of seismic impulses from airgun arrays and sighting rates of cetaceans, but they caution that their analysis was so crude that it was unable to detect changes in distribution of less than 100 km. Their study exemplifies the critical point that a reported lack of an effect must carefully specify the statistical power of a study to detect specific effects. Other studies sighting marine mammals closer to sound sources have found avoidance at several hundred to thousands of meters (e.g., Goold, 1996). And some studies have shown no displacements. Ringed seals *(Phoca hispida)* near an artificial-island drilling site were monitored before and during development of the site. Although in-air and underwater sound was audible to the seals for up to 5 km, there was no change in their density in that area between breeding seasons before and breeding seasons after development began (Moulton et al., 2003).

The last few decades have seen the development of experiments designed to study the causal relationship between exposure to sound and behavior. As Tyack et al. (2004) discuss, these controlled-exposure experiments (CEEs) are similar to playback experiments that are commonly used to study animal communication. The primary difference is that CEEs carefully titrate the acoustic exposure required to elicit a specific behavioral reaction. In few studies have responses of marine mammals been related to levels of anthropogenic sounds. Playback of sounds associated with oil-industry activities indicated a clear relationship between the received-sound pressure level and the probability that migrating gray whales will deviate from their migration path. For continuous sounds, about 50% of the whales avoided exposure to about 120 dB rms re 1 µPa; for short impulses from airguns (about 0.01 sec every 10 sec), 50% avoidance occurred at about 170 dB re 1 µPa (Malme et al., 1983, 1984; Tyack, 1998; airgun levels are average pulse pressures). Tyack and Clark (1998) replicated the earlier experiments of Malme and colleagues by using Surveillance Towed Array Sensor System-Low Frequency Active (SURTASS-LFA) sonar sounds transmitted for 42 sec every 6 min and found that course deflection occurred when the received levels were about 140 dB rms re 1 µPa. Not only was there a steady increase in avoidance with increasing received level of each stimulus type, but there also was a clear pattern in which higher levels were required to achieve the same avoidance when signals were of shorter duration and lower duty cycle. Similar relationships between temporary threshold shift (TTS) and duration of the sound have been shown in laboratory studies (see below in the discussion of physiological effects).

Other CEEs have found a relationship between received level of sound and probability of some responses and less relationship for others. In a playback experiment involving the SURTASS-LFA sound and singing humpback whales *(Megaptera novaeangliae)*, Fristrup et al. (2003) analyzed 378 songs recorded before, during, and after playback. They found that the songs of the humpback whales were longer when the playback was louder (they could not determine received level at the whale). Miller et al. (2000) followed 16 singers during 18 of the same playbacks. During 18 playbacks, nine of the whales stopped singing. Of the nine, four stopped when they joined with another whale (a normal baseline behavior), so, there were five cessations of song potentially in response to the sonar (although whales stop singing without joining even under baseline conditions). The received levels measured next to the whales were 120-150 dB rms re 1 µPa, and there was no relationship between received level and the probability of cessation of singing. For six whales in which at least one complete song was recorded during the playback, the songs were an average of 29% longer. Miller et al. (2000) did not find a significant increase in song length with received playback level, probably because their study was less powerful than that of a larger sample analyzed in Fristrup et al. (2003). A similar CEE with responses of right whales to three 2-min stimuli, 60% duty cycles, and energy of 500-4,500 Hz showed no relation between probability or strength of response and received level, which was 133-148 dB rms (Nowacek et al., 2003), but this result is also limited by the small sample.

Both observational studies and CEEs demonstrate that behavioral context can have a substantial effect on relationships of acoustic dosage to behavioral response. For example, Tyack and Clark (1998) report that the avoidance reaction found when the SURTASS-LFA sound source was placed in the middle of the migration path apparently disappeared when the sound source was placed just offshore of the main migration path, even if the whales passed close to the source. On a larger scale, beluga whales in the Canadian high arctic show intense and prolonged reactions to the propulsion sounds of icebreakers (Cosens and Dueck, 1988; Finley et al., 1990), whereas beluga whales in Bristol Bay, Alaska, continued to feed when surrounded by fishing vessels and resisted dispersal even when purposely harassed by motorboats (Fish and Vania, 1971). This context specificity of behavioral reactions to sound raises questions about the ecological validity of extrapolating data from captive animals to the wild.

The behavioral responses of marine mammals to acoustic stimuli vary widely, depending on the species, the context, the properties of the stimuli, and prior exposure of the animals (Wartzok et al., 2004). Species variation in auditory processing is so important that a distinction should certainly be made between taxonomic groups that have widely different hearing and sensitivity frequencies. For example, pinnipeds have lower maximal frequency of hearing and maximal sensitivity of hearing than odontocetes (toothed whales). They typically have a high-frequency cutoff in their underwater hearing between 30 and 60 kHz, and maximal sensitivity of about 60 dB re 1 µPa, and odontocetes have best frequency of hearing between 80 and 150 kHz and maximum sensitivity between 40-50 dB. Therefore, odontocetes can hear over a wider frequency range and have keener hearing than pinnipeds, so they could potentially be affected by a wider variety of sounds. Little is known about the frequency range of hearing and sensitivity of some marine mammal taxa, such as baleen whales, but several attempts have been made to divide marine mammals into functional categories on the basis of hearing (e.g., Ketten, 1994).

As mentioned above, some of the variation in responses between species or individuals may stem from differences in audition. Not only do different species have different hearing capabilities but there is considerable variation in hearing among conspecifics. One of the most predictable patterns in mammals involves age-related hearing loss, which particularly affects high frequencies and is more common in males than females (Willott et al., 2001).

Auditory processing is less likely than behavior to differ between captive and wild animals, and captive data on behavioral reactions closely linked to audition may be relevant to other settings. For example, Schlundt et al. (2000) noted disturbance reactions of captive bottlenose dolphins *(Tursiops truncatus)* and beluga whales during TTS experiments. The behavioral reactions involved avoidance of the source, refusal of participation in the test, aggressive threats, or attacks on the equipment. Finneran and Schlundt (2004) showed that the probability of those reactions increased with increasing received level from 160 to 200 dB rms re 1 µPa at 1m except for low-frequency (400-Hz) stimuli near the low-frequency boundary of auditory sensitivity. The kinds of reactions observed and how they scale with intense exposures near the level that provoked TTS suggest that the signals were perceived as annoyingly loud.

Some of the variation in responses to sound may stem from experience. There are several well-known mechanisms by which an animal modifies its responses to a sound stimulus, depending upon reinforcement correlated with exposure. The response of animals to an innocuous stimulus often wanes after repeated exposure—a process called habituation. The National Research Council (NRC, 1993) recommended studies on habituation of marine mammals to repeated human-made sounds. In one of few experimental studies of habituation in marine mammals, Cox et al. (2001) showed that porpoises tended to avoid at a distance of 208 m upon initial exposure to a 10-kHz pinger with a source level of 132 dB peak to peak re 1 µPa at 1m. This avoidance distance dropped by 50% within 4 days, and sightings within 125 m equaled control values within 10-11 days. The pingers are used on nets to prevent porpoises from becoming entangled in them, so evaluations of their effectiveness must take habituation into account.

Kastelein et al. (1997) report that a captive harbor porpoise *(Phocoena phocoena)* avoided exposure to high-frequency pingers with source levels of 103-117 dB rms re 1 µPa at 1m and received levels of 78-90 dB rms re 1 µPa. When exposed to a source with a level of 158 dB rms re 1 µPa at 1m, the porpoise swam as far away as possible in the enclosure and made shallow rapid dives. Those results combine with the results of Cox et al. (2001) to suggest that porpoises react to sound at much lower levels than the captive delphinids studied by Finneran and Schlundt (2004). However, the context of the captive studies was quite different: the dolphins and belugas studied by Finneran and Schlundt were being rewarded for submitting to exposure to intense sounds, whereas the porpoise was not being rewarded for remaining in the sound field.

If an animal in captivity or the wild is conditioned to associate a sound with a food reward, it may become more tolerant of the sound and may become sensitized and use the sound as a cue for foraging. Several large-scale studies have shown that the distribution of feeding baleen whales correlates with prey but not with loud sonar or industrial activities (Croll et al., 2001); but the studies were unable to test for potentially more subtle effects on feeding, such as reduced prey capture per unit effort and reduced time engaged in feeding.

Some of the strongest reactions of marine mammals to human-generated noise may occur when the sound happens to match their general template for predator sounds. The risk-benefit relationship is very different for predator defense and foraging. An animal may lose a meal if it fails to recognize a foraging opportunity, but it may die if it fails to detect predators. Animals do not have the luxury of learning to detect predators through experience with them. Deecke et al. (2002) showed that harbor seals responded strongly to playbacks of the calls of mammal-eating killer whales and unfamiliar fish-eating killer whales but not to familiar calls of local fish-eating killer whales. That suggests that, like birds studied with visual models of predators (Schleidt, 1961a; 1961b), these animals inherit diffuse templates for predators. They initially respond to any stimulus similar to the predator template but learn through habituation to cease responding to harmless variants of the general predator image.

It would make sense for animals to show strong reactions to novel sounds that fit within the predator template, whatever the received level. Indeed, the behavioral reactions of belugas to ice breaker noise match the local Inuit description of their responses to killer whales, a dangerous predator. Some of those strong reactions to novel sounds, such as the responses of diving right whales to an artificial alarm stimulus as reported by Nowacek et al. (2003), might be expected to habituate if the stimuli are distinguishable from real predators and are not associated with aversive effects. In fact, the only right whale subject not to respond was the last of six whales tested, and it may have heard the stimulus up to five times before. Beluga whales that fled icebreaker noise at received levels of 94-105 dB rms re 1 µPa returned in 1-2 days to the area where received icebreaker noise was 120 dB rms re 1 µPa (Finley et al., 1990). In contrast, Kastak and Schusterman (1996) reported that a captive elephant seal not only did not habituate but was sensitized to a broadband pulsed stimulus somewhat similar to killer whale echolocation clicks even though nothing dangerous or aversive was associated with the noise.

The low sound levels that stimulate intense responses of Arctic beluga whales (Frost et al., 1984; LGL and Greeneridge, 1986; Cosens and Dueck, 1988) contrast sharply with the high levels required to evoke responses in captive beluga whales (Finneran and Schlundt, 2004). This difference highlights that there are likely to be several kinds of response, depending on whether the animal is captive and whether the noise resembles that of a known predator. Annoyance responses may require levels of sound well above levels that may stimulate strong antipredator responses. If animals in the wild hear a sound that matches their auditory template for a predator, they may avoid exposures to sound levels much lower than those required to elicit the disturbance responses observed by Finneran and Schlundt (2004). If learning can modify the predator template, as suggested by

Deecke et al. (2002), it is essential to conduct studies of behavioral responses of animals to human-made stimuli in habitats resembling those encountered by wild populations.

An important property of most anthropogenic sound is that high-intensity levels are typically confined to the immediate location of the sound source (an exception is high-intensity, low-frequency sound), so any effects caused by exposure to high levels are reduced as animals move away from the source. However, high-intensity low-frequency sound travels well enough underwater that animals can detect signals at ranges of tens to hundreds of kilometers from the source. If, as in the case of Arctic belugas hearing icebreaker noise, exposure to low received levels can still trigger an intense response, a few sources may affect a large fraction of a population.

Even in the absence of a strong response, low received levels of sound can affect a large fraction of a population if the sound results in a masking of normal stimuli. Marine mammals show exquisite adaptations to overcome masking, but they may not be effective in the presence of pervasive anthropogenic sounds (reviewed in NRC, 2003b; Wartzok et al., 2004).

PHYSIOLOGICAL RESPONSES TO ACOUSTIC STIMULI

Auditory Damage

Most discussions of physiological effects of noise have centered on the auditory system. Audition has evolved for sensitivity to sound, so it is likely to be the physiological system most sensitive to acoustic stimuli that are within the frequency range of hearing. When the mammalian auditory system is exposed to a high level of sounds for a specific duration, the hair cells in the cochlea begin to fatigue and do not immediately return to their normal shape. When the hair cells fatigue in that way, the animal's hearing becomes less sensitive. If the exposure is below some critical energy flux density limit, the hair cells will eventually return to their normal shape; the hearing loss will be temporary, and the effect is termed a *temporary threshold shift* in hearing sensitivity, or TTS. If the sound exposure exceeds a higher limit, the hair cells in the cochlea become permanently damaged and will eventually die; the hearing loss will be permanent, and the effect is termed a *permanent threshold shift* in sensitivity, or PTS. TTS and PTS limits vary among individuals in a population, so they need to be characterized statistically. A relationship between the TTS limit and the PTS limit has been determined for laboratory animals; the appropriateness of extrapolating of such a relationship to marine mammals is untested.

A major recommendation of the National Research Council 1994 report supported the development of TTS studies in marine mammals. Since then, TTS experiments have been conducted in two species of odontocetes (*Tursiops truncatus* and *Delphinapterus leucas*) with both behavioral and electrophysiological techniques (Finneran et al., 2000; Schlundt et al., 2000; Nachtigall et al., 2003, 2004) and three species of pinnipeds (*Phoca vitulina, Zalophus californianus,* and *Mirounga augustirostris*) with behavioral techniques (Kastak et al., 1999; Finneran et al., 2002). Those experiments were conducted at three centers for research on marine mammals that have facilities to hold their own animals: the Hawaii Institute of Marine Biology of the University of Hawaii, Long Marine Laboratory of the University of California, Santa Cruz, and the Space and Naval Warfare Systems Center (SPAWAR) of the US Navy in San Diego. The scientists at the Hawaii Institute of Marine Biology used continuous random noise with a bandwidth slightly greater than 1 octave as the fatiguing stimulus and both behavioral and electrophysiological techniques to measure TTS in the bottlenose dolphin. The fatiguing stimulus had a broadband received level of 179 dB rms re 1 µPa, which was about 99 dB above the animal's pure-tone threshold of 80 dB at the test-tone frequency of 7.5 kHz (Nachtigall et al., 2003). Exposure to 50 min of the fatiguing stimulus resulted in a TTS of 2-18 dB. Recovery from the TTS occurred within 20 minutes after the cessation of the fatiguing stimulus. More recent studies (Nachtigall et al., 2004) that used an auditory brainstem response showed a TTS of 5-8 dB in response to 30 minutes of a 160-dB rms re 1 µPa fatiguing stimulus. Although the intensity of the fatiguing stimulus fell rapidly above 11 kHz, the greatest TTS was shown at 16 kHz. This pattern of TTS being more prominent at a frequency above the frequency of the fatiguing stimulus matches results for humans (Ward, 1963). The recovery occurred at 1.5 dB per doubling time with complete recovery within 45 min. The 1.5 dB recovery per doubling time was also found for recovery from the more intense 179 dB fatiguing stimulus used in the earlier study (Nachtigall et al., 2003). Researchers at Long Marine Laboratory used continuous random noise of 1-octave bandwidth as the fatiguing stimulus and a behavioral technique to measure TTS in the harbor seal *(Phoca vitulina),* California sea lion *(Zalophus californianus),* and elephant seal *(Mirounga augustirostris).* They exposed the subjects to 20-22 min of the fatiguing stimulus and found that it only had to be 60-75 dB above the hearing threshold to induce a TTS of 4-5 dB for test signals at frequencies between 100 Hz and 2 kHz. Threshold measurements conducted 24 hours after the cessation of the fatiguing stimulus indicated complete recovery from the TTS (Kastak et al., 1999). Researchers at SPAWAR used impulse sounds from a seismic watergun as the fatiguing stimulus and a behavioral technique to measure the TTS (Finneran et al., 2002). The fatiguing stimulus had a variable duration of about 1 ms, peak pressure of 160 kPa, a sound pressure of 226 dB peak-to-peak re 1 µPa at 1m, and an energy flux density of 186 dB re 1 µPa^{2}s, which produced a TTS of 7 and 6 dB at 0.4 and 30 kHz respectively in beluga whales but not at the other tested frequency of 4 kHz. In dolphins, no TTS could be demonstrated at 0.4, 4 or 30 kHz in spite of raising the fatiguing stimulus to its maximum intensity of 228 dB (Finneran et al., 2002). Each of these experiments used different durations of fatiguing stimuli. When the sound pressure required to produce a TTS is plotted against the duration of the stimulus for all these experiments, the result is a line with a slope of -3 dB per doubling of stimulus duration, that is, a line showing that the TTS occurred at about an equal energy in all cases tested to date.

Changes in hearing threshold, even TTSs, have the potential to affect population vital rates through increased predation or decreased foraging sources of individual animals that experience a TTS as they use sound for these tasks. A TTS also has the potential to decrease the range over which socially significant communication takes place, for example, between competing males, between males and females during mating season, and between mothers and offspring. Unless a critical opportunity is available only during a narrow time window, the potential effects on vital rates are important only if exposures and any resulting TTS are prolonged. In spite of the importance of sound for marine mammals, there is considerable variability in hearing sensitivity within a species, and there is evidence of age-related hearing loss.

Nonauditory Effects of Sound

Resonance Effects

A marine mammal has many airspaces and gas-filled tissues that could theoretically be driven into resonance by impinging acoustic energy. The lungs, air-filled sinuses that include those of the middle ears, and in the intestines, where there can be small gas bubbles, are among the areas that may be susceptible to resonance induced by acoustic sources. However, there were no published measurements of resonance in a marine mammal until the work of Finneran (2003), who measured the resonance of the lungs of a bottlenose dolphin and a beluga whale. Before Finneran's work, most studies of acoustic damage in marine mammals concentrated on the effects of shock waves, including blast-related phenomena.

Finneran (2003) used a backscatter technique to measure the resonance of the lungs of a 280-kg bottlenose dolphin and a 540-kg beluga whale. He obtained resonance frequencies of 30 Hz for the larger white whale and 36 Hz for the bottlenose dolphin. However, the resonance was highly damped and far less intense than predicted by a free-standing bubble model. The lungs experience a symmetric expansion and contraction when ensonified. How intensely a structure resonates at its resonant frequency can be quantified, and is represented by Q. The higher the Q, the more resonant the structure. The Q values measured in marine mammals are low. The Q of the lungs of the beluga whale was found to be 2.5, and of the bottlenose dolphin 3.1. Those Q values suggest a broad resonance property that is highly damped. Apparently, the tissue and other mass surrounding the lungs dampen the susceptibility of the lungs and probably other structures to resonate intensely.

Although other gas-filled structures will resonate at different frequencies, the probable low Q values, as in Finneran's study, suggest that resonance of air spaces is not likely to lead to detrimental physiological effects on marine mammals. That was also the conclusion of a panel of experts convened by National Oceanic and Atmospheric Administration (NOAA) Fisheries (NOAA, 2002).

Rectified Diffusion

Rectified diffusion is a physical phenomenon that leads to the growth of microscopic bubble nuclei in the presence of high-intensity sound. It has been demonstrated only in laboratory preparations, but it is theoretically possible that exposure to high-intensity sound could enhance bubble growth in humans and marine mammals (Crum and Mao, 1996). Rectified diffusion might be a possible mechanism of nonauditory acoustic trauma in human divers and marine mammals, in that bubbles in tissue or blood can lead to injury or death. Calculations by Crum and Mao (1996) suggest that, given a modest degree of nitrogen (N_2) supersaturation of biological tissues (for example, between 100% and 200%), the growth of normally stabilized nuclei would be unlikely to occur at sound pressures below 190 dB rms re 1 µPa. However, at sound pressures above 210 dB, significant bubble growth could occur. As nitrogen supersaturation increases, the exposure threshold of activation should decrease, and the growth rate of bubbles should increase.

Houser et al. (2001) modeled the accumulation of N_2 in the muscle of diving cetaceans on the basis of dive profiles of deep-diving odontocetes and data on N_2 accumulation previously measured in the muscle of diving bottlenose dolphins (Ridgway and Howard, 1979). The model necessarily assumed that N_2 kinetics were the same between species and that lung collapse occurred at 70 m—a prediction made by Ridgway and Howard for bottlenose dolphins. The conclusions of the model were that slow deep-diving cetaceans (diving beyond the depth of lung collapse), which had few extended surface intervals, would accumulate the greatest amount of N_2 in their tissues while diving. The slower the dive in water shallower than lung collapse, the longer the time the animal experiences pressure that drives the accumulation of gas in the tissues; short surface intervals between deep dives would limit the time the animal has to clear accumulated N_2 from its body.

The magnitude of tissue N_2 supersaturation—and thus the possibility of rectified diffusion—depends on dive behavior as described above. Records of dive behavior of beaked whales—Cuvier's beaked whale (*Ziphius cavirostris*) and Blainville's beaked whale (*Mesoplodon densirostris*)—presented at a recent workshop (Marine Mammal Commission, 2004) indicate that these animals have long deep dives followed by a short surfacing and than a series of shallow dives primarily within the region in which gas exchange occurs in the lung. The short surfacings and the repeated "bounce" dives near the surface could lead to high tissue N_2 pressure and the possibility of bubble formation. Those are the predominant species of beaked whales that have stranded in association with naval sonar activity, although other beaked whale species have also been involved.

Evidence of deleterious bubble formation in diving cetaceans and the putative causative mechanisms (acoustically and behaviorally mediated) remain open to debate. Jepson et al. (2003) conducted necropsies of stranded cetaceans and reported on signs of bubble-related injury, but their interpretation has been challenged (Piantadosi and Thalmann, 2004). No experimental evidence has been collected on the feasibility of the putative mechanisms of bubble formation in breath-hold divers. More research is needed to understand the role of rectified diffusion in marine mammals, but our current understanding suggests that it would be relevant only for animals exposed to sound substantially above 180 dB re 1 µPa, which is already considered by regulators to be a threshold for risk of other forms of injury.

PROGRESS ON EARLIER NATIONAL RESEARCH COUNCIL RECOMMENDATIONS

Three previous National Research Council reports recommended research to resolve critical uncertainties about the effects of noise on marine mammals (1994, 2000, 2003b). All three highlight the need for research in behavioral ecology, auditory physiology and anatomy, nonauditory effects of sound, effects of sound on prey of marine mammals, and development of new techniques. The 2003 report also recommended research on sources and modeling of ocean noise. Some of the recommendations have led to research that has greatly reduced the data gap. For example, the 1994 and 2000 reports recommended experiments to determine acoustic exposures that would lead to temporary shifts in the threshold of hearing of marine mammals. In the last decade, several laboratories have succeeded in conducting the experiments; as a result, the uncertainty involved in modeling the noise exposures that start to cause physiological effects on hearing has been reduced.

Progress has also been made on the recommendation with respect to development of new technology. For example, the 1994 report recommended the development of tags to record physiology, behavior, location, and sound exposure. In the last decade, tags have been developed to record all but physiological characteristics (Johnson and Tyack, 2003).

For many of the other research recommendations, research is being conducted, but progress has been slow enough over the last decade to argue for the establishment of a targeted research program. The 2000 and 2003 reports recommended better coordination between federal regulatory agencies and science-funding agencies to develop a multidisciplinary research program. It was recommended that the research program operate like that of the National Science Foundation and the Office of Naval Research, issuing targeted requests for proposals and judging the quality of proposals with peer review. Although some progress has been made, it is worth reiterating that progress on critical research requires that the federal government develop and fund a dedicated research program.

3

How to Get from Acoustic Disturbance to Population Effects

The committee developed a conceptual model, named the Population Consequences of Acoustic Disturbance (PCAD) model as a framework to break the overwhelmingly difficult task of tracing acoustic stimuli to population effects into several manageable stages (Figure 3-1). The PCAD model was created as a first attempt to trace acoustic disturbance through the life history of a marine mammal and then to determine the consequences for the population. The model also serves as a framework for identifying existing data and data gaps. The model was distributed to workshop participants (Appendix D) before the workshop, discussed during the workshop, and, with the input of workshop participants, refined afterwards. The model involves five levels of variables that are related by four transfer functions. The first transfer function relates acoustic stimuli to behavioral responses. The second expresses behavioral disruption in terms of effects on critical life functions, such as feeding and breeding. The third integrates these functional outcomes of responses over daily and seasonal cycles, to link them to vital rates (see Figure 3-1) in life history. The fourth transfer function relates changes in the vital rates of individuals to population effects. Current data are insufficient to allow the PCAD model to serve as more than a conceptual model, so the listing of data at the first three levels—involving sound characteristics, behaviors, and life functions—is exemplary rather than all-inclusive. The relationship between vital rates and population effects is well defined, but the specification of relevant population effects involves policy decisions as well as scientific judgments.

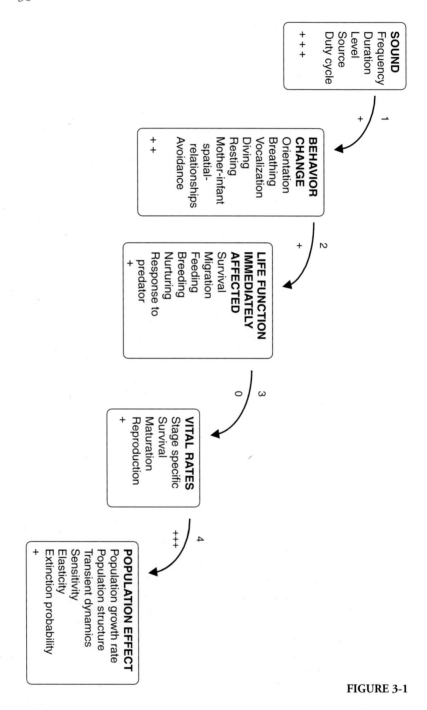

FIGURE 3-1

The bottom entry in each data level presents an indicator of how readily available or easily obtainable the critical data are.

Although it would be unrealistic to expect to acquire the data necessary to run such a model for all species of marine mammals, it will be important to model a representative sample of pinnipeds, baleen whales, and odontocetes with different hearing ranges and hearing anatomies (Ketten, 1994). The recently discovered particular sensitivity of beaked whales to mid-frequency tactical sonar (D'Amico, 1998; Evans and England, 2001; Jepson et al., 2003) demonstrates the necessity for both care and comprehensiveness in the selection of representative species. The 2000 National Research Council report provided a reasonable selection of species, sound types, and behavioral responses that could be used in the PCAD model (Box 3-1).

FINDING: Prior National Research Council reports (NRC, 1994, 2000, 2003b) identified high-priority subjects of research. The model proposed here requires the data and understanding that will become available on the fulfillment of the earlier National Research Council high-priority research recommendations.

RECOMMENDATION 1: The high-priority research identified by the National Research Council (1994, 2000, 2003b) should be completed. That research is essential for the model proposed in this report.

FIGURE 3-1 The conceptual Population Consequences of Acoustic Disturbance model describes several stages required to relate acoustic disturbance to effects on a marine mammal population. Five groups of variables are of interest, and transfer functions specify the relationships between the variables listed, for example, how sounds of a given frequency affect the vocalization rate of a given species of marine mammal under specified conditions. Each box lists variables with observable features (sound, behavior change, life function affected, vital rates, and population effect). In most cases, the causal mechanisms of responses are not known. For example, survival is included as one of the life functions that could be affected to account for such situations as the beaked whale strandings, in which it is generally agreed that exposure to sound leads to death. The causal steps between reception of sound and death are by no means known or agreed on, but the result is clear. The "+" signs at the bottoms of the boxes indicate how well the variables can be measured. The indicators between boxes show how well the "black box" nature of the transfer functions is understood; these indicators scale from "+++" (well known and easily observed) to "0" (unknown).

BOX 3-1
High-Priority Research for Whales and Seals Recommended by National Research Council in 2000

To move beyond the requirement for extensive study of each sound source and each area in which it may be operated (NRC [2000] recommended that), a coordinated plan should be developed to explore how sound characteristics affect the responses of a representative set of marine mammal species in several biological contexts (e.g., feeding, migrating, and breeding). Research should be focused on studies of representative species using standard signal types, measuring a standard set of biological parameters, based on hearing type (Ketten, 1994), taxonomic group, and behavioral ecology (at least one species per group). This could allow the development of mathematical models that predict the levels and types of noise that pose a risk of injury or behavioral disturbance to marine mammals. Such models could be used to predict in multidimensional space where temporary threshold shift (TTS) is likely (a "TTS potential region") and TTS can be used as a threshold of potential risk of injury to the auditory systems. This coordinated plan can be used to set priorities for research required to determine measures of behavioral disruption for different species groups. Observations should include both trained (where possible) and wild animals (with attention to ecological validity). The results of such research could provide the necessary background for future environmental impact statements, regulations, and permitting processes.

Groupings of Species Estimated to Have Similar Sensitivity to Sound. Research and observations should be conducted on at least one species in each of the following seven groups:

1. Sperm whales (*Physeter macrocephalus*; not to include other physterids)
2. Baleen whales
3. Beaked whales
4. Pygmy (*Kogia breviceps*) and dwarf sperm whales (*Kogia sima*) and porpoises (high-frequency [greater than 100 kHz] narrowband sonar signals)

5. Delphinids (dolphins, white whales [*Delphinapterus leucas*], narwhals [*Monodon monoceros*], killer whales)
6. Phocids (true seals) and walruses
7. Otariids (eared seals and sea lions)

Signal Type. Standardized analytic signals should be developed for testing with individuals of the preceding seven species groups. These signals should emulate the signals used for human activities in the ocean, including impulse and continuous sources.

1. Impulse—airguns, explosions, sparkers, some types of sonar
2. Transient—frequency-modulated (low-frequency [LFA], other sonars, animal sounds), amplitude-modulated (animal sounds, ship passage), broadband (sonar)
3. Continuous—frequency-modulated, amplitude-modulated (drilling rigs), broadband (ship noise)

Biological Parameters to Measure. When testing representative species, several different biological parameters should be measured as a basis for future regulations and individual permitting decisions. These parameters include the following:

- Mortality
- TTS at signal frequency and other frequencies
- Injury—permanent threshold shifts
- Level B harassment
- Avoidance
- Masking (temporal and spectral)
- Absolute sensitivity
- Temporal integration function
- Nonauditory biological effects
- Biologically significant behaviors with the potential to change demographic parameters such as mortality and reproduction.

Modified from NRC (2000).

All the transfer functions in the PCAD model may vary depending on the season and the species, location, age, and sex of the animal. Other external factors may also modulate the responses and effects. For example, behavioral responses that would be insignificant in a normal year may become biologically significant during an El Niño year. Behavioral responses, on individual and population scales, may differ between a stable population near environmental carrying capacity and a severely depleted population. Those types of modulations are considered in the model in two primary, but not exclusive, categories: time and energy budgets and homeostasis and allostasis.

The first stages in the PCAD model are relatively clear. In general, the characteristics of the sounds can be measured accurately. In some cases, the behavioral responses of the animals can be measured as well. Although mechanistic models that relate sounds to behavior are unavailable, such an understanding is not essential for management use of this model if the behavioral changes can be measured and predicted.

Dose-response studies are a good way to quantify the first transfer function, relationship of sounds to behavioral responses. For marine mammals, data are available on only a few sounds and a few behaviors in a few species. Observational and correlational studies can provide trend data, and expert-opinion modeling can provide at least a "lookup" table to serve as a surrogate for this transfer function (Andelman et al., 2001). NOAA Fisheries has convened a panel of acoustic experts to survey the literature on mammalian hearing and the effects of noise and to draft noise-exposure criteria for five functional hearing groups of marine mammals (low-, middle-, and high-frequency cetaceans, pinnipeds in water, and pinnipeds in air) exposed to four sound types (single and multiple pulses and nonpulses). The criteria are based on individual sound-exposure events in which either the sound pressure (rms or peak) or the energy flux density exceeds one of two impact levels. The impact levels are tissue injury and behavioral disruption. Thus, the full matrix has 80 threshold criteria—the product of five animal groups, four sound types, two exposure metrics and two impact levels. The NOAA Fisheries panel has presented some portions of the criteria but has yet to complete a final draft. Some key elements of the criteria remain undetermined, particularly with regard to behavioral disturbance.

The second stage of the PCAD model attempts to evaluate how changes in behavior may affect life functions that are widely recognized as critical to population dynamics. With the exception of direct impact on life, the exact relationship of these functions to life-history characteristics is largely unknown. Furthermore, the impacts of sound on these functions through behavior will be difficult to measure.

Time-scale integration is important in identifying impacts and determining relationships between changes in behavior patterns and resulting changes in life functions. Because most marine mammals have a diurnal cycle of activities, integration of short-term functional consequences over a duration of at least 24 hours may be appropriate and could be studied by using behavioral observations or electronic tagging methods. In addition, however, most marine mammals also have strong seasonal variations in behavior and physiology. As more data are accumulated, daily functional consequences might be summed over each season in relation to the expected duration of exposure to the specific sound of interest to evaluate daily and seasonal effects in a particular species.

The final stages of the model relate changes in life functions in individuals to impacts at the population level. There are at least two components to these final stages. The first is the most difficult—relating changes in life functions to changes in vital rates of individuals. If the link from life functions to age-specific vital rates is known, changes in population dynamics can be explored by using demographic analyses. Current demographic theory provides the capability to deal with vital rates not only on the basis of age but also in terms of biologically defined stages that reflect developmental, behavioral, or physiological properties of individuals.

A critical question is what population consequences should be identified as significant. The measure of population performance must integrate survival and reproduction across the lifespan. It should have implications for recovery, persistence, and extinction of populations. The most thoroughly investigated index is the population growth rate (modified by such adjectives as potential, intrinsic, and asymptotic). Demographic theory provides tools that explicitly link changes in the life cycle to changes in population growth rate. That makes demographic models a powerful tool for placing bounds on likely effects, for exploring the quantitative implications of hypothesized interactions, and for synthesizing what is currently known. Establishing acceptable population effects is a management question that has already received a good deal of attention. One example used for protecting marine mammals involves setting the potential number of marine mammals that can be removed from the population without endangering the population. The management criteria of this Potential Biological Removal (PBR) model (Taylor et al., 2000; also see Chapter 4) are:

- Healthy populations will remain above the optimal sustainable population (OSP) numbers, as defined in the Marine Mammal Protection Act (MMPA), over the next 20 years.
- Recovering populations will reach OSP numbers after 100 years.
- The recovery of populations at high risk will not be delayed in reaching OSP numbers by more than 10% beyond the predicted time if there is no human-induced mortality.

The amount of information needed to map from sound to its population consequences is truly enormous. The PCAD diagram should be thought of not as the blueprint for an eventual universal model, but as a framework that clarifies where different kinds of information fit in and identifies processes that need study. Years of work will be required to accumulate data and develop models for the transfer functions between behavior and life functions, and between life functions and the vital rates. This report is essentially a status report and a roadmap for this critical long-term project of turning a conceptual model into a predictive model useful for science-based management of marine mammals and their exposure to sound. In the interim, techniques must be developed to use current information more effectively in making science-based management decisions. After discussion of the PCAD model, we propose (in Chapter 4) a means to achieve better management over a shorter timeline.

FINDING: A conceptual model, such as the PCAD model, is useful for clarifying the complex problem of acoustic-disturbance effects on marine mammal populations. Such a model can be used as a framework for identifying the cause-effect relationships necessary for determining consequences of disturbances. Data to complete this exercise are lacking and must be pursued from every available source.

> **RECOMMENDATION 2: A conceptual model, such as that described in this report, should be developed more fully to help to assess impacts of acoustic disturbance on marine mammal populations. Development of such a model will allow sensitivity analysis that can be used to focus, stimulate, and direct research on appropriate transfer functions.**

In addition to research studies designed to evaluate reactions of marine mammals to noise, limited information is available from monitoring programs that are required of some activities that might "take" small numbers of marine mammals as defined in the MMPA. The incidental-harassment authorizations issued by the US government often contain the requirement for the operator to implement a program to monitor effects on marine mammals. For activities that produce intense noise, such as seismic surveys, the monitoring requirement often involves sighting animals from the vessel that is introducing the noise. Sighting surveys are also required by the United Kingdom and have been summarized in reports that identify avoidance reactions to seismic surveys (Stone, 2001, 2003). Few of those studies measured the acoustic stimulus from the activity as heard by the animal, and they typically scored easy-to-observe changes in behavior, such as respiration rate, time on the surface, duration of dive, change in swimming speed or direction, avoidance behavior, and aerial display. However, if those short-term measures are selected purely for ease of observation, it will often be difficult to link the responses to the functional categories described in the PCAD model, a link that is essential for extrapolating short-term measures to long-term effects that would alter some life function of an individual animal. Federal regulators for the last several decades have required monitoring programs instead of targeted research on the assumption that monitoring would detect developing problems. Monitoring programs, as implemented, have seldom provided the relevant data, suggesting that regulators and the regulated community should consider altering the balance of resources that they provide for monitoring versus research.

The impact of a behavioral reaction to sound depends on the number of animals affected in a population and on the duration and intensity of the reaction. The impact of avoidance reactions depends in the short term on the percentage of habitat reduction and on the ease with which animals can move to and use alternative habitat. Determining overall impact on the population requires estimation of

- The range of the impact of individual sources in time and space.
- The number of animals and the fraction of the population affected.
- The total impact of all sources deployed.
- The intensity of reaction of each animal.
- The duration of the impact on each animal.

The presence of anthropogenic sound sources could have minimal effects on a healthy population that can relocate with minimal effort or could be devastating to a small population that is living on the edge of its capabilities to survive where the sources affect its entire habitat (Box 3-2).

One of the few subjects of research that provide predictive models with connections from behavioral ecology through physiology to demography is how animals obtain and use energy. Behavioral ecologists have developed models to predict how animals maximize energy intake per unit of time as they forage (Stevens and Krebs, 1987). Physiologists and physiological ecologists have developed models to predict the baseline metabolic rates of animals and the metabolic costs of various activities. If a foraging animal takes in more energy than it uses for metabolism, it builds up an energy surplus that can be used for growth or reproduction. All large mammals have an initial period of sexual immaturity in which most surplus energy

---

BOX 3-2
Special Considerations for Endangered Populations

The effects of seismic surveys on western gray whales (*Eschrichtius robustus*) off Sakhalin Island, Russia, illustrate the potential for anthropogenic sound to have a severe impact on a marine mammal population. The western gray whale is critically endangered, numbering about 100, and depends on the northeastern Sakhalin Island feeding ground for most of its food intake. Weller et al. (2002) and Johnson (2002) report displacement of some whales during seismic surveys in 2001, and Johnson (2002) reports observations of gray whale behavior suggesting that they may have spent more time traveling and less time feeding during exposure to seismic signals, but aerial observations of feeding plumes were unable to detect any changes in feeding activity related to seismic activity. Disruption of feeding in preferred areas, especially in a small population in which many females (with and without calves) are already in poor condition and have long intervals between calf production (Brownell and Weller, 2002), could have major impacts on individual whales, their reproductive success, and even the survival of this critically endangered population (Weller et al., 2002). Observed changes in the distribution of individuals of this highly endangered population could be critical; deciphering their impact will require more detailed studies of prey distribution, foraging ecology, and energetics of these whales.

reserves go to growth. The timing of the transition to sexual maturity is affected by the need to have grown to a particular point and the need to have amassed sufficient energy reserves to support the energy cost of the transition. Once a female is mature, the timing between her ovulations, the probability of successfully carrying a fetus to term, and the interval between offspring are all affected adversely by lack of energy resources. Those characteristics are all used directly in demographic models to estimate the reproductive rate of the population. During periods of starvation, the probability of survival may be affected if the animal's metabolism exceeds energy intake for long periods. When foraging is not adequate, animals may abandon their young to conserve energy for their own survival. Limited energy resources may also make animals more vulnerable to other stressors (as discussed below in the section on allostasis). The various models that link foraging behavior, energy reserves, reproduction, and survival offer great promise for our proposed PCAD model, but more effort will be required to link the different submodels. The strength of research on energy budgets suggests that studies of effects of noise on foraging animals should focus on effects of disruption of time-energy budgets.

As noted earlier, repeated reports of unusual mass strandings of Blainville's and Cuvier's beaked whales show a correlation with naval maneuvers. The locations of whales with respect to the ships operating the sonars are unknown and cannot be reconstructed. However, the timing and spatial extent of the strandings suggest a possible risk of stranding for whales exposed to noise as low as 160 dB re 1 µPa. Current data on physiological or behavioral effects of well-studied marine mammals would not have suggested such a risk to poorly known beaked whales. The recent cases of the association of beaked whale strandings with naval sonar stimulated a review of prior records of beaked whale mass stranding events (Brownell et al., 2004; Taylor et al., 2004). This historical review indicated that mass strandings of beaked whales have occurred primarily subsequent to the introduction of mid-frequency tactical sonar in the early 1960s. However, caution must be exercised in these post hoc correlational studies. For example, when the radius of potential correlation extends to 500 km, as was the case with the strandings of *Z. cavirostris* and seismic in the Galapagos (Taylor et al., 2004), the potential for false positives increases proportionally. Therefore, there is a critical need for carefully designed and executed epidemiological studies to find potential problems as well as toxicological studies to evaluate precise dose-response relationships.

CURRENT DATA COLLECTION EFFORTS

In addition to basic research conducted primarily through universities and published in a host of peer-reviewed scientific journals, many data on marine mammals are gathered to fulfill regulatory requirements. For example, every permit application under the MMPA or the Endangered Species Act requires the applicant to provide a summary of the best available information on the status of the affected species or stock and on factors that affect the status. Permits for scientific research also contain many relevant data with respect to the habitat, behavior, physiology, or demography of the animals. A condition of many permits is the requirement to monitor the animals sighted, the time, location and oceanographic conditions of the observations, and the responses of the animals to the permitted activities. Federal agencies with responsibility for managing marine mammal populations conduct intramural research that often ends up as unpublished reports that contain valuable information. For example, NOAA Fisheries conducts surveys for assessing the status of marine mammal stocks. The agency publishes regular stock-assessment reports, but the sighting data would be extremely valuable for other purposes, such as predicting the species and number of animals that might be exposed to sound in a particular place and at a particular time.

Information from all these sources, with appropriate indicators of the sources, should be integrated into a common database. Peer-reviewed data and interpretation should be given the highest quality indicator. Other data sources should have appropriate quality indicators assigned. To facilitate the integration of data from many sources, federal agencies should establish standard data-reporting formats to be used in permit applications, permit reports, and research sponsored by other entities in fulfillment of permit requirements. Some federal support has been provided to begin the development of such integrated databases. Examples of such support are the Office of Naval Research Effects of Sound on the Marine Environment project and the Marine Resource Assessments by the Commander in Chiefs, U.S. Atlantic and Pacific Fleets.

FINDING: A wealth of data on marine mammals is collected in compliance with federal regulatory requirements. Such data are not collected in a manner that allows easy access or use beyond the original intent of their collection (such as permit issuance). A data-coordination effort could improve our knowledge of marine mammal distribution, behavior, and population size; improve and standardize data used for regulatory processes; and greatly reduce the effort required of applicants for permits or authorization.

RECOMMENDATION 3: To assist in the development of the conceptual model, a centralized database of marine mammal sightings and their responses to anthropogenic sound in the ocean should be developed and should include

- Published peer-reviewed papers in the scientific literature.
- Government technical reports.
- Data submitted to NOAA Fisheries and the US Fish and Wildlife Service in permit applications.
- Data submitted by industry to the Minerals Management Service for regulating off-shore hydrocarbon exploration and production.
- All relevant data accumulated by all federal agencies in the course of their research and operational activities, including monitoring.

To facilitate the integration of data from the various sources, federal agencies need to develop standardized data-reporting formats. Survey data should include locations where marine mammals were detected and the track lines when personnel were monitoring for marine mammals, regardless of whether any were sighted. All data entered into such an integrated database must be coded as to quality, and peer-reviewed data and interpretations should be rated highest.

DATA NEEDED TO DETERMINE PHYSIOLOGICAL RESPONSES TO ACOUSTIC STIMULI

Immediate behavioral responses are the easiest to observe, but the population consequences of sound will be modulated through physiological responses. The ear is the body structure most sensitive to acoustic input and is the site at which acoustic energy in the frequency range of hearing is most likely to have direct physiological effects. This report reiterates the recommendations of the 1994 and 2000 National Research Council reports to acquire more data on assessments of hearing characteristics such as absolute sensitivity, masking, temporary threshold shifts, and temporal integration, and on the evaluation of behavior during exposure. However, the long-term effects of noise exposure on individuals can be best determined through the physiological integration that occurs and can be observed as indicators of cumulative stress.

Physiological Stress Effects

Anthropogenic sound is a potential source of stress in marine mammals, and it has been shown to increase blood pressure and catecholamine and cortisol concentrations in humans (Evans et al., 1995; Evans et al., 2001; Ising and Kruppa, 2004). Biomedical research on stress has provided a theoretical framework that can help scientists to conceptualize and ultimately measure the cumulative impact of multiple stressors on individual animals (McEwen and Stellar, 1993; Seeman et al., 2001). Application of the concepts, theories, and techniques to marine mammals could accelerate our understanding of the physiological effects of noise and other stressors on them.

Historically, the term *stress* has been used to refer to several concepts, including noxious stimuli, the physiological and behavioral coping responses of organisms to noxious stimuli, and the pathological states that result when the coping responses can no longer restore the body to a normal condition. Several attempts have been made to provide a less ambiguous terminology. For example, Romero (2004) refers to the three concepts listed above as *stressors*, the *stress response*, and *chronic stress*, respectively, and this terminology is used productively in the physiological and behavioral literature. An alternative terminology, which we will consider in some detail because of its conceptual integration with energy budgets and life-history events, has been offered by McEwen and Wingfield (2003). It focuses on the concept of allostatic load, which was adapted from the cardiovascular field and was introduced for more broad application and developed by McEwen and colleagues (McEwen and Stellar, 1993).

McEwen and Wingfield (2003) propose four terms—*allostasis, allostatic state, allostatic load,* and *allostatic overload,* that can be considered in relation to the life cycle and energy budget of any species (Figure 3-2). Although energy is a convenient currency to consider for illustrative purposes, it could be replaced in Figure 3-2 by any other resource vital to survival, such as a particular vitamin or mineral. *Allostasis* refers to the physiological and behavioral mechanisms used by an organism to support homeostasis (the

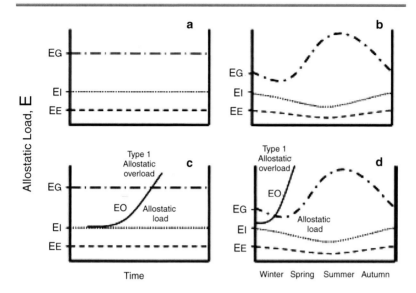

FIGURE 3-2 A framework for modeling energy requirements (E) of organisms during their life cycle. This energy requirement, E, includes all potential nutritional requirements, including energy itself. These separate and distinct requirements are represented more generally here for convenience, although essential components of nutrition could also be modeled separately. EE represents the energy required for basic homeostasis. EI represents the extra energy required for the organism to find, process, and assimilate food in ideal conditions. EG represents the amount of energy (in food) available in the environment (from Wingfield et al., 1998; Wingfield and Ramenofsky, 1999). (a) The three are represented as straight lines when environmental and physiological states do not vary over time. (b) The changes in the quantities have been adjusted to represent probable seasonal changes. EG would be expected to rise dramatically in spring and summer, when primary productivity is high, and then decline through autumn and winter, when primary productivity is low. In this scenario, EE would be lowest in summer, when ambient temperatures are highest. EI should be fairly constant (under ideal conditions) and should vary in parallel with EE. (c) EO represents additional costs incurred after a perturbation (such as a storm or anthropogenic disturbance) that increases costs above EE + EI. It represents the energy required to find food, process it, and assimilate nutrients in nonideal conditions. Allostatic load (see text) increases as EO persists. If EO exceeds EG, type 1 allostatic overload begins, resulting in an increase in plasma glucocorticosteroids. That usually triggers an emergency condition that results in altered physiology and behavior that reduces allostatic overload. (d) In more naturally fluctuating conditions, type 1 allostatic overload may occur first or more rapidly if a perturbation occurs during a season when the conditions are worse. If insults are permanent—such as those caused by even mild or moderate but persistent disturbances

caption continues

stability of the physiological systems that maintain life) in the face of normal and relatively predictable life-history events, such as migration, mating, rearing young, and seasonal changes in resource availability; unpredictable events, such as decreases in oceanic productivity and increases in human disturbance; and more permanent handicaps, such as injuries, parasites, and contaminant loads. Examples of allostatic responses are the physiological changes that occur in lactating female mammals (Bauman, 2000) and the changes in metabolism, muscle morphology, and behavior that occur in migrating birds (Kuenzel et al., 1999).

An "allostatic state" is a sustained imbalance in the physiological mediators, such as various hormones, that integrate behavioral and physiological responses to changing environmental conditions. An allostatic state can be maintained for some time if environmental resources are sufficient. However, the cumulative result of an organism's allostatic state is its "allostatic load." The usual allostatic load results from the organism's need to obtain enough food to survive plus any extra energy required for normal seasonal activities, such as migrating, molting, mating, and lactating. Animals can adapt to the extra demands within limits. However, if resources in the environment are insufficient (Figure 3-2d) or if other challenges—such as disease, human disturbance, or stressful social interactions—increase the allostatic load, the animal can no longer cope and will develop serious pathological conditions or die.

The concept of allostasis makes it clear that the effect of any given stressor will be contingent on multiple factors, including species, sex, nutritional and reproductive condition, and any other stressors currently affecting an animal. The closer an animal is to the condition of allostatic overload when subjected to an additional stressor, the more likely it is that

---

FIGURE 3-2 caption continued

(e.g., Creel et al., 2002 for effects of snowmobile activity on wolves and elk), abnormally high densities of animals, increased pollutants (Porter et al., 1999), disadvantageous social status in some terrestrial species (Goymann and Wingfield, 2004; Sands and Creel, 2004), or individual differences in emotional or other vulnerable states (Sapolsky, 1994 for baboons; Cavigelli and McClintock, 2003 for rats)—overload will occur in most seasons and will be triggered readily even in seasons or conditions that are otherwise benign.

SOURCE: Adapted from McEwen and Wingfield, 2003.

the additional stressor will have a deleterious impact. That is, the effect of a stressor often depends heavily on the context in which it occurs. The importance of context has also been shown by laboratory experiments that demonstrate that uncontrollable and unpredictable stimuli cause a greater stress response than controllable and predictable stimuli. For example, when two rats are given similar electrical shocks but only one can press a lever that terminates the shock for both, the rat that can terminate the shock has a dramatically lower hormonal response to the shock than the one that has no control over the length of the shock (Weiss, 1968).

Consideration of energy needs can also provide clues to the conditions in which marine mammals may be most likely to suffer allostatic overload. The following account of marine mammal energetics follows the recent review by Boyd (2002), who built on earlier reviews for pinnipeds (Lavigne et al., 1982; Costa, 1991, 1993) and cetaceans (Lockyer, 1981). Different species have different energy requirements and appear to balance their energy budgets by developing body sizes and life histories that match the distribution and abundance of their food. As body size increases, the period over which an animal must balance its energy budget lengthens. For example, the great baleen whales probably balance their energy budget on a 1-year cycle. They typically migrate to high latitudes during the summer to feed on krill or other seasonally abundant resources and store enough energy in the form of blubber for them to be able to fast for the rest of the year and reproduce in warmer but less productive tropical waters. Smaller species, such as most of the odontocetes (dolphins and porpoises), must balance their energy budgets on much shorter periods—months to days. Thus, energy considerations suggest that sound disturbance could severely affect the energy budget of baleen whales if it displaced them from their feeding grounds for a substantial fraction of the feeding season but would be less likely to have a serious effect on energy needs if it occurred in other circumstances, such as during migration, and merely displaced them temporarily from their normal migratory path.

The diverse lactation strategies of female pinnipeds provide a particularly good illustration of the relationships between body size, energetics, and behavior. Lactating pinnipeds nourish their pups from a food supply that may be near or very distant from the rookeries where they give birth. If sufficient food is available near the birth site for mothers to balance their own energy budget and provide for the pups, mothers make foraging trips during lactation. That strategy is followed by most of the otariids (fur seals and sea lions), which are relatively small for marine mammals (Costa,

1993). Larger species, such as most phocids (true seals), can forage over larger areas and use more dispersed prey resources. They can feed on lower-quality prey and need greater rates of prey consumption, but they can use a patchier prey distribution. For example, elephant seals feed thousands of kilometers from the sites where they give birth and, like the baleen whales, store enough energy in the form of fat to be able to fast while lactating (Costa et al., 1986; Boness and Bowen, 1996). Phocids appear to switch during lactation from foraging to fasting at a body mass of about 100 kg; harbor seals (80-100 kg) forage during lactation whereas gray seals (*Halichoerus grypus*, 130-180 kg) fast (Costa, 1991; Boyd, 2002).

The reproductive success of small pinnipeds that make repeated short-distance foraging trips during lactation is severely affected if they are unable to acquire normal amounts of prey. Evidence of that is provided by El Niño events, which occur at irregular intervals that tend to range between 2 and 7 years and result in greatly decreased productivity in the eastern tropical Pacific and greatly reduce the survival of pinniped pups (Trillmich and Ono, 1991). For example, during the 1982 El Niño, pup production was normal, but none of the pups survived the first 5 months after birth. In 1983, pup production was only 11% of normal, but survival of the pups returned to normal rates (Trillmich and Dellinger, 1991). Thus, energy considerations suggest that small otariid species could be affected rather quickly by anthropogenic noise close to their rookeries if it interrupted normal foraging whereas larger species that were not foraging during lactation would be more likely to meet their and their pups' energy needs in the presence of a similar disturbance.

The physiological stress response is highly conserved and similar across vertebrate taxa (Wingfield and Romero, 2001). As an integrator of stresses, neuroendocrinological responses include both direct and indirect effects of noise exposure. Physiological responses to stressors are initiated by activation of the hypothalamic-pituitary-adrenal axis, which results in the release of catecholamines and stress hormones, such as glucocorticoids, from the adrenal glands (McEwen, 2000). Because the extent of the stress response often correlates with the general health of an animal, measuring the response can serve as a general indicator of the current condition of an animal, reflecting its health, its energy allocation, and the effect of human disturbances on it. The promise of applying this approach in the field is illustrated by recent research on marine iguanas *(Amblyrhynchus cristatus)* in the Galapagos Islands (Romero and Wikelski, 2001, 2002). During El Niño years, iguanas had higher baseline corticosteroid concentrations during famines. Handling of the iguanas also resulted in higher stress-induced corticosteroid concentrations than in normal years. Stress-induced corticosteroid concentrations in animals were good predictors of whether they would survive an El Niño event (Romero and Wikelski, 2001). Measurement of corticosteroid stress responses also showed that apparently low levels of oil contamination caused a strong hormonal stress response in iguanas. That response accurately predicted higher mortality over the next year among iguanas on oil-contaminated islands than on uncontaminated islands (Romero and Wikelski, 2002). A growing body of literature on terrestrial mammals has demonstrated sensitivity of glucocorticoids to sudden natural social stressors (e.g., Alberts et al., 1992 for wild baboons), to persistent natural stressors (e.g., Sapolsky, 1994), and to anthropogenic stressors (e.g., Creel et al., 2002 for wolves and elk).

Glucocorticoids may be part of the mechanisms by which behavioral effects are translated into altered rates of reproduction and mortality, and in other instances they will at least be indicators if not major players in the cascade of effects leading from behavior to survival and reproduction. As indicated above, it will be feasible in some cases to obtain fairly convincing evidence of the behavior-demography relationships with or without the physiological links between the two; but in most others, our greatest power will come from documenting behavior-glucocorticoid relationships in some studies and glucocorticoid-survival or glucocorticoid-reproduction relationships in others, as suggested by a number of studies already cited. Examples of an emerging picture of behavior-demography or behavior-glucocorticoid relationships from one of the best-studied wild large mammal species have been found in baboons (Box 3-3).

Physiological indicators of body condition and of pregnancy can be obtained from serum. Serum sampling of glucorticoid concentrations can also be used to obtain a physiological stress measure if the sample can be obtained before the stress of capture and sampling changes hormone concentrations in the blood. The maximal allowable time from capture to blood sampling is 2-3 min for small birds or rodents and 10-15 min for large monkeys. Determining the time for various marine mammals will identify the extent to which this technique can be applied usefully, at least in situations where capture for blood sampling is feasible.

In most cases, capture of marine mammals for blood sampling will be impossible. Instead, techniques will need to be developed to allow unrestrained blood sampling. Hill (1986) developed a package that could be attached to a freely diving Weddell seal and could take blood samples on

BOX 3-3
Behavior, Physiology, and Demography in Baboons

In baboons, a number of behavioral differences have been associated with altered demographics. Reduced travel time to foraging sites leads to a net positive increase in energy balance (Muruthi et al., 1991) and presumably thereby to the observed decreased age of maturation (Altmann et al., 1993), doubling of reproduction (halving of interbirth interval), and increasing offspring survival (J. Altmann, Department of Ecology and Evolutionary Biology, Princeton University, Princeton, NJ, unpublished data, 2004; S.C. Alberts, Department of Biology, Duke University, Durham, NC, unpublished data, 2004) despite increased rates of aggression (Altmann and Muruthi, 1988). Daughters and sons of low-status females mature later (Altmann et al., 1988; Alberts and Altmann, 1995). Larger baboon social groups are associated with decreased reproductive rates of lower-status females (Altmann and Alberts, 2003). Infants of females that are more social have higher survival (Silk et al., 2003). Effects of chronic or sudden behavioral differences on stress hormones have also been demonstrated in baboon studies. Among baboon males, either social status or degree of sociality affects glucocorticoid concentrations (Sapolsky et al., 1997), as does social style or recent winning or losing of fights (Sapolsky, 1994). Sudden social disruption by immigration of an aggressive male leads to high glucocorticoids in both sexes and in the aggressive immigrant itself (Alberts et al., 1992). Despite that body of data, however, studies linking small chronic differences in glucocorticoids to vital rates in such a large mammal are only now possible and are being conducted thanks to the recent techniques in fecal steroid analysis.

---

a programmed schedule. More recently, sophisticated data logger tags have been attached to marine mammals to study their responses to anthropogenic sounds (Burgess, 2001; Johnson and Tyack, 2003). Data logging packages could be modified to incorporate blood sampling during controlled-exposure experiments (CEEs). Initial studies would likely need to be conducted on Weddell seals constrained to returning to an isolated hole to breathe. Eventually, the packages would benefit from the ability to take blood samples on a controlled basis and stabilize hormones for later analysis or to conduct "on-board" blood-chemistry analysis to record responses of animals in situations less unusual than that of the Weddell seal.

Totally noninvasive, hands-off techniques of sampling glucocorticoids and other steroid hormone metabolites (such as estrogens, testosterone metabolites, and progestins) through collection of feces or urine are increasingly used for terrestrial mammals in situations or with species that make capture or any disruption to behavior intolerable (e.g., Wasser et al., 2000). The feasibility of feces collection from some marine mammals in the ocean has been demonstrated (Rolland et al., 2004); validation and calibration of the assays should have high priority (Buchanan and Goldsmith, 2004; Hunt et al., 2004). Preliminary studies measuring glucocorticoids in hair samples and up-regulation of stress-induced proteins in skin samples merit further development. Concentrations of fecal progestins are increasingly used in research and conservation for assessing pregnancy in terrestrial mammals. Application to marine mammals to evaluate pregnancy rates and fetal or early infant loss may be relatively straightforward (Larson et al., 2003) when the requisite samples can be obtained.

FINDING: Measurements of glucocorticoids and other steroid hormone metabolites in terrestrial vertebrates have proved to be good indicators of pregnancy, allostatic overload, and mortality risk posed by current and new stressors.

FINDING: Continued development of more-sophisticated data logger tags is necessary to advance the study of marine mammal responses to anthropogenic sounds. Data logging packages should be modified to incorporate blood sampling during controlled-exposure experiments (CEEs).

> **RECOMMENDATION 4:** The use of glucocorticoid and other serum hormone concentrations to assess stress should be developed, validated, and calibrated for various marine mammal species and age-sex classes and conditions. Dose-Response curves for those indicators as a function of sound characteristics need to be established. Development of a sampling package that could take blood samples on a controlled basis and stabilize hormones for later analysis or process samples "on-board" for corticosteroids at various stages of a CEE would be invaluable for determining the stress that the sound is producing. The use of fecal sampling to measure condition or stress needs

to be investigated further and developed. Research efforts should seek to determine whether reliable long-term stress indicators exist and, if so, whether they can be used to differentiate between noise-induced stress and other sources of stress in representative marine mammal species (this recommendation was also made in NRC, 2003b).

Toxicology

The concept of allostasis provides a framework for understanding how anthropogenic noises that at first appear insignificant could, with repeated exposure or in combination with other stressors, compromise an animal's survival and reproduction. Recent research in toxicology has provided cautionary examples of how the combined actions of apparently safe individual factors can have serious unforeseen consequences. For example, a mixture of several agrochemicals at concentrations commonly found in groundwater across the United States affected immune, endocrine, and nervous system function in wild deer mice (*Peromyscus maniculatus*) and outbred white mice when consumed for 14-103 days (Porter et al., 1999). In this 5-year study with a full factorial design, numerous deleterious changes occurred in response to mixtures of aldicarb (an insecticide), atrazine (a herbicide), and nitrate (a fertilizer) at low concentrations, but the changes were rarely seen when the compounds were tested individually at the same concentrations. In another study, a commercial herbicide containing a mixture of 2,4-dichlorophenoxyacetic acid (2,4-D), mecoprop, dicamba, and several inert ingredients led to a U-shaped dose-response curve for litter size in mice; the lowest dosages of the mixture caused the greatest decrease in the number of live pups born (Cavieres et al., 2002). Such studies demonstrate that multiple stressors can interact in complex and unforeseen ways to produce adverse effects on living organisms.

DATA NEEDED TO DESCRIBE MARINE MAMMAL POPULATIONS

To understand the behavioral effect that a sound may have in a given place and at a given time, it is necessary to be able to answer the following questions:

- What species are present?
- What is their distribution?
- What are their grouping patterns?
- What activities are they engaged in?
- How is each activity disrupted by sound?

NOAA Fisheries has collected and analyzed data on sightings of marine mammals to assess the status of different populations, and extensive sighting data are available from other sources, but the data are not available in a form that allows the prediction of the number of animals likely to be exposed to a sound in a given place and at a given time. Grouping patterns are important because if animals live in groups an average density will not yield a correct probability of the number of animals exposed.

Even fewer data are available on how marine mammals use different areas. That data gap could be addressed by completing basic behavioral ecological studies of marine mammals in the wild. To understand the biological significance of behavioral disruption, a greatly accelerated program is needed for studying the behavior and ecology of marine mammals in the wild, with a focus on how variation in behavior may affect probabilities of survival, growth, and reproduction in different ecological settings. The first recommendations for research in the 1994 and 2000 National Research Council reports were to study the behavior of marine mammals in the wild. Ten years after the 1994 report, a major increase in support of research to fill this critical data gap is still needed. The urgency of a research program in this field is highlighted by the PCAD model.

INDIVIDUALS TO POPULATIONS: USING MODELS TO IMPROVE UNDERSTANDING

In the PCAD model, there are at least some data that link sounds to behavioral responses of individuals. The connection between individuals and the population is much more speculative. There are good reasons for this lack of data. Most effects on life functions are separated in time and space from the immediate behavioral responses to sounds. Thus, if later observations identified life-function activities outside the normal range, it would be difficult to relate them to prior exposure to sound. Furthermore, our current understanding of the behaviors associated with most life functions is incomplete. For example, we do not yet fully understand normal ranges of the behaviors, so effects may not be detected even if they are observable. As noted previously, there is almost no understanding of how changes in any of the life functions lead to changes in vital rates.

The only way to build a bridge from the individual to a population is modeling of some kind. No single model will serve the purpose, but a number of modeling exercises could help to integrate what is known tactically (in the short term) and to structure strategic research in the longer term. We consider here the types of modeling that might prove helpful and the expectations for each.

Uses of Models: Prediction and Exploration

The use of models for prediction is most successful when a well-established understanding of the processes and a good database for parameterizing the model exist. With respect to linking individual to population effects in marine mammals, both understanding and data are lacking (Figure 3-3, Area 4). Predictive modeling to determine the population effects of noise on marine mammals is therefore not now an option.

The determination of an appropriate modeling technique depends on the information and understanding available (Starfield and Bleloch, 1991). A schematic representation can be used to describe possible approaches (Figure 3-3). Area 1 is the region of good data but little understanding; statistical tools are applicable and can be used to perform an exploratory data analysis (sensu Tukey, 1977) to search for patterns and relationships. Area 3 is the region of good data and good understanding where predictive modeling has the best chance of success; well-established paradigms and modeling approaches can be used with confidence and are backed by experience and theory.

If either data quantity or quality is poor, a modeler is restricted to Areas 2 and 4, referred to as "data-limited." In Area 2, there is good understanding of the processes and structure of the problem; in Area 4, that understanding is weak. Marine mammal data are still sparse, so this report is concerned mainly with Areas 2 and 4. Issues in these two areas present the modeler with two daunting challenges:

- Despite the lack of data and understanding, a management or policy decision must be made. How can modelers help to make the best scientific decisions under these circumstances?

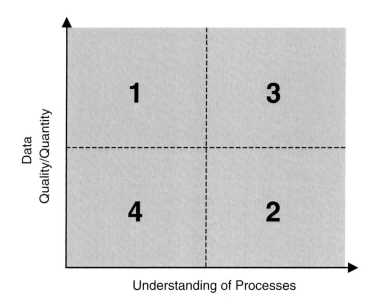

FIGURE 3-3 Classification of modeling problems.
SOURCE: Holling, 1978.

- How can models be used to exploit available data to improve understanding and, in turn, identify data that are critically needed? In other words, how do we progress from Area 4 toward Area 3?

Scientists and modelers are often uncomfortable dealing with these issues. Some believe that modeling should be confined to Area 3; others push ahead and try to use predictive models in an area where prediction is, to say the least, risky. Starfield and Bleloch (1991) suggest that Areas 2 and 4 require a different paradigm in which models are used tentatively to explore alternative hypotheses, speculate on possible outcomes given whatever data are available, and then cautiously reach some conclusions—even if they are only conclusions about future research needs. The way that they propose using models in Areas 2 and 4 is philosophically akin to Tukey's exploratory data analysis; that is why the term *exploratory modeling* is used.

Exploratory models can be used as tools for synthesizing what is known, explaining what may be happening, or perhaps guiding research or management. In all cases, if modeling is to serve a useful function, it is essential that the purpose or objective of each modeling exercise is clearly defined, the model is focused ruthlessly on the objectives, and all involved with the modeling exercises have a pragmatic appreciation of the power (or lack thereof) of whatever modeling paradigm is developed. These are some of the objectives for a suite of modeling exercises:

- *Objective 1: To bound the problem or look for significant thresholds.* In the introduction, it was stated that it is not clear whether noise has a second-order effect on populations or whether what has so far been observed is only the tip of the iceberg. Models could help to categorize the likely effect of specified noise doses on different populations. That type of modeling exercise would be useful even if it produced the limited result that "Dose X is unlikely to have a measurable effect on a population with these characteristics, but it could have a measurable effect on a population with those characteristics."
- *Objective 2: To speculate on the likely outcome of hypothesized interactions.* The objective is to take a word model (such as "disruptions of courtship in species X will have a significant impact on the recovery of the species") and tease out the implications quantitatively. The modeling would perforce be speculative, but there is value to exploring explicitly which assumptions and which sets of parameter values support the hypothesis. To quote Samuel Johnson, "That, sir, is the good of counting. It brings everything to a certainty which before floated in the mind indefinitely."
- *Objective 3: To synthesize and organize what is currently known.* For example, we know that responses can be situation-specific. It has already been noted that the responses of migrating gray whales depend on whether a low-frequency active source is in the migratory path or a few kilometers seaward of the migratory path even though the received levels were similar (Tyack and Clark, 2000) and that the responses of beluga whales in the high arctic to the initial seasonal exposure to an icebreaker are stronger and more prolonged (Cosens and Dueck, 1988; Finley et al., 1990) than the responses of beluga in the same region to the icebreaker a few days later (Finley et al., 1990) and the responses of Bristol Bay, Alaska, beluga to

direct harassment attempts (Fish and Vania, 1971). It is likely that more is known than appears to be known—synthesis will produce more than the sum of the parts. There are three related objectives:

(1) To focus the attention of disciplinary scientists on how their knowledge and data can be combined or otherwise used to address the problem.
(2) To identify gaps in data and knowledge and explore what one minimally needs to do to fill the gaps.
(3) To provide guidelines for data collection and monitoring.

- *Objective 4: To develop a conceptual framework for management guidelines.* Models can be used to organize and improve management guidelines, such as described for the Potential Biological Removal management regime described in Chapter 4.

A number of alternative modeling paradigms and constructs could fit with some of those objectives. For example, the age-structured demographic models (Caswell, 1989) usually used for predictive modeling could be used in an exploratory way to help to bound the problem and establish thresholds for different species. It is difficult to be specific about suitable paradigms or the design of a model until the precise objectives of an exploratory modeling exercise are spelled out, but two additional potential approaches are offered: individual-based models (IBMs) and categorical or qualitative models. In the next sections of this report we describe three modeling approaches and two additional tools that might prove helpful.

Demographic Models

The most well-developed and widely used approach to population modeling is that of age-structured demographic analysis. A demographic model is one that categorizes individuals into groups based on biological characteristics relevant to their survival and reproduction. In classical demography those groups were based on age (and implicitly on sex), but it is now known that other criteria, such as maturity, reproductive status, physiological condition, and spatial location may be more important (e.g., Caswell, 2001). Stage-structured models are most commonly expressed as population projection matrices, which may include environmental stochasticity, demographic stochasticity, density-dependence, and spatial structure (Caswell, 2001).

Demographic models can be analyzed to obtain measures of population growth and structure, probabilities of extinction or quasi-extinction, and other measures of population performance. They employ a well-developed perturbation theory that permits calculation of the effect of changes in the vital rates on those measures of performance; this makes them particularly suitable for the exploration of thresholds and the effects of interactions. Matrix population models have a well-developed connection with statistical methods for parameter estimation, especially from observations of known individuals (e.g., Nichols et al., 1992; Fujiwara and Caswell, 2001, 2002a,b; Caswell and Fujiwara, 2004). These methods can incorporate measurements of individual animal condition into estimates of the vital rates. Recent research has explored Bayesian methods for parameter estimation in these models (Gross et al., 2002); such methods are particularly suitable for analysis of uncertainty.

Matrix models have been used for demographic analysis of killer whales (Brault and Caswell, 1993), humpback whales (Barlow and Clapham, 1997), right whales (Fujiwara and Caswell, 2001; Fujiwara, 2002), and harbor porpoises (Caswell et al., 1998) as well as various species of seals (e.g., Heide-Jorgensen et al., 1992; York, 1994; Kokko et al., 1997; Lalas and Bradshaw, 2003).

Although demographic models could be used to make predictions, their most common use is to explore the consequences of various biological processes in the face of unknown data. In two cases, the California condor (Mertz, 1971) and the Everglades kite (Nichols et al., 1980), only the most fragmentary data were available—both studies used demographic models to place bounds on population growth, to speculate on the outcome of hypothesized interactions, and to synthesize sparse data. More recent examples of exploratory use of demographic models include the exploration of management strategies for sea turtles (Crouse et al., 1987), the exploration of bycatch effects in harbor porpoise (Caswell et al., 1998), and exploration of research priorities for the sooty shearwater (Hunter et al., 2000).

Individual-Based Models

In an Individual-Based Model (IBM), the computer program is designed to simulate virtual individuals in a population, often from birth to death. Each individual carries a set of attributes or markers that describe the state of the individual. They can include demographic factors, such as age and sex; energy factors, such as weight, stomach fullness, and diet composition; location descriptors, such as latitude and longitude; and behavioral descriptors, such as reproductive status, dive intensity, and dominance role. Such programs as Tagging of Pacific Pelagics (Block et al., 2003) provide data on movement patterns in relation to oceanographic features and seasonal patterns of movement essential for constructing a valid IBM for these species. IBMs have been constructed for species in a variety of habitats (Grimm, 1999).

For example, an IBM has been designed to compare the effects of alternative trophy-hunting strategies (Whitman et al., 2004). It describes each male lion (at any time step) in terms of his age, social status (cub, nomad, or pride lion), associates (like-aged cubs, fellow nomads in a nomadic group, and fellow males in a pride coalition), and spatial position (which pride a cub is born into, which territories a pride coalition controls and patrols, and which territories a nomadic group is temporarily visiting). Those attributes enable one to simulate such processes as competition between neighboring pride males, territorial battles between resident pride males and visiting nomads, and infanticide when pride coalitions are replaced—all essential to an understanding of how trophy hunting might affect the size and structure of a lion population. Some other examples are the modeling of deer and Florida panther (*Puma concolor coryi*) populations in the Florida Everglades (Abbott et al., 1997) and of walleye pollock (*Theragra chalcogramma*) in the western Gulf of Alaska (Hermann et al., 2001).

IBMs can be used for purposes similar to those of structured demographic models and can also directly address questions about the interaction between, for example, behavior of animals in relation to a source and the resulting acoustic exposure, behavior and reproduction, or behavior and growth. They offer a direct venue for considering the effects of noise on marine mammal individuals and populations. They can accommodate the kinds of data that are now becoming available on the relationships between behavior and acoustic exposure in a direct and comprehensible fashion. For example, the Acoustic Integration Model (AIM; Frankel et al., 2002) models the location and dive behavior of simulated marine mammals swimming near a modeled acoustic source. An acoustic-propagation model is used to predict the exposure of the simulated animals and can program different response strategies of the animals for the simulated source. It has been used to predict the exposure of animals with different response patterns to sources with different modes of operation, monitoring, and mitigation; and it can help in selecting alternatives that minimize effects on marine mammals while maximizing operational effectiveness of the source.

Categorical or Qualitative Models

The links or transfer functions between changes in the behavior of individuals, effects on life functions, and effects on vital rates (survival and reproduction) of a population in the conceptual model (Figure 3-1) have been identified as ones on which there is little information. However, some progress might be made by combining whatever is known with an understanding of the behaviors and pressure points in different species to derive a qualitative ranking of the strength of a link. An example of behaviors and pressure points would be a marine mammal with an "income" breeding strategy (Costa, 1993) of intensively nursing newborn pups in bouts separated by extensive time at sea to replenish reserves. It can be argued that a reduction in the feeding success of mothers during that period will have a more severe effect on pup survival than an equivalent reduction in feeding success in a capital breeder (an animal that relies on stored energy to survive the breeding season).

A categorical or qualitative model would characterize effects in such terms as low, moderate, and severe. Such a model may separate the consequences of an effect from the probability that it will occur. It could be developed with a combination of available information on marine mammals, information on comparable nonmarine mammals selected on the basis of life-history scaling or body-size scaling, first principles, and expert opinion (Morgan and Henrion, 1990; Goodwin and Wright, 1991; Meyer and Booker, 1991; Anderson, 1998; Andelman et al., 2001).

The Scientific Committee on Antarctic Research (SCAR, 2004) created a series of risk-assessment matrices for different acoustic sources in Antarctic waters. The cells of a likelihood-consequences matrix indicated whether there was a potential risk to an individual or the population. One conclusion of this analysis was that the risks associated with the use of most scientific acoustic equipment in the Antarctic were less than or comparable with the risk associated simply with the passage of the research ship through Antarctic waters.

Categorical or qualitative models might serve two purposes: to create a structure for encouraging biologists to make the best determinations they can and to explore the feasibility of developing tactical management strategies akin to the PBR model (see Chapter 4). Essential components of such a model would be estimates of the reliability of every categorization in the model and explanations of how each categorization was reached. The models would provide a structure for further refinement and, like the proposed IBM and demographic modeling exercises, help to identify gaps in knowledge. The key point to make is that modeling exercises like this can lead to robust management approaches, as the PBR model demonstrates, even when knowledge is incomplete.

Expert Opinion

Data on many links in the chain from acoustic stimuli to population effects on marine mammal populations are sparse or lacking. Therefore, regulators such as NOAA Fisheries and the US Fish and Wildlife Service (FWS) may often find it necessary to rely on expert opinion regarding the probable effects of specific activities until more data accumulate. Although the use of expert opinion does not necessarily produce an accurate result (experts can be wrong, especially when data are lacking), it does provide a structured, well-documented basis for decision-making that often withstands legal scrutiny. Precedents for the use of expert opinion to evaluate risk in a conservation context are provided by the US Department of Agriculture Forest Service's extensive reliance on expert opinion for population-viability assessments under the National Forest Management Act (Andelman et al., 2001) and FWS's increasing use of expert opinion for making listing decisions under the Endangered Species Act (ESA; J. Cochrane, US Fish and Wildlife Service, personal communication, 2004). Because eliciting and using expert opinion are complex tasks beset with pitfalls for the inexperienced, any use of expert opinion should follow established procedures detailed in the substantial scientific literature on the subject (Morgan and Henrion, 1990; Goodwin and Wright, 1991; Meyer and Booker, 1991; Anderson, 1998; Andelman et al., 2001) to avoid bias and increase credibility.

Risk Assessment

Evaluating the effects of noise on marine mammal populations is a problem in risk assessment. Previous National Research Council reports have considered the general process of risk assessment by the federal government (NRC, 1983) and risk assessment in relation to contaminants and human health (NRC, 1993). Uncertainty is always a prominent feature of risk assessment, and uncertainty regarding the probable effects of human activities on marine mammals is not limited to the effects of noise but rather is a pervasive problem, which can be addressed using population models (Caswell et al., 1998; Ralls and Taylor, 2000).

Risk assessment can be combined with decision analysis to make management decisions in the face of uncertainty (Harwood, 2000, 2002). The general approach is discussed in detail with respect to making decisions under the ESA in an earlier Research Council report (NRC, 1995). There are two main categories of errors in judging the effects of human activities on natural resources: we may conclude that a risk is great when it is not, which leads to overprotection and unnecessary economic loss, or we may conclude that a risk is small when it is not, which leads to underprotection and avoidable loss of a valued resource. It is impossible to minimize simultaneously the probability of making those two types of errors, and common statistical practices of hypothesis-testing may lead to a systematic bias against the welfare of species or populations that are in need of protective action (NRC, 1995, Chapter 8). Analyzing risks with the framework of decision analysis increases the probability that all types of errors and their consequences are adequately considered.

Advances in technology have enabled the use of computer-intensive methods in risk assessment (e.g., Slooten et al., 2000; Taylor et al., 2000). If relevant data on marine mammals are lacking, this kind of simulation approach can benefit from the use of data on other species selected on the basis of life history, ecology, or body size (e.g., Caswell et al., 1998). Bayesian decision theory, which allows choices among more than two decisions, offers many advantages and is increasingly recommended for use in risk assessment related to natural-resources management (Ludwig, 1996; Taylor et al., 1996; Wade, 2000).

FINDING: Focused effort is needed on a modeling exercise that should include demographic models, IBMs, and categorical modeling. Such an effort should start with, and calibrate against, expert opinion and should incorporate such characteristics as

- An aim to pull together what is known—in different ways, from different disciplines—and to assess both the importance and the degrees of uncertainty associated with the information.

- The use of tactical models, with the objective of probing how successfully current knowledge could be applied.
- The use of structured models to test hypotheses.
- The use of models to identify crucial gaps in knowledge. (A gap in knowledge is not just something we do not know; it is something we do not know and need to know if we are to meet our objectives.)
- An aim to encourage interdisciplinary synthesis and provide a structure for it.
- The requirement that all modeling efforts be explicit about uncertainty and its consequences.
- A similar requirement that all models clearly state their limited purpose and that both their strengths and their shortcomings be evaluated.
- A risk assessment for the species being modeled if the model is to be used for management decisions.

RECOMMENDATION 5: Several marine mammal species for which there are good long-term demographic and behavioral data on individuals should be selected as targets of an intensive exploratory modeling effort that would develop a series of individual-based models and stage- or age-structured demographic models for the species as appropriate. NOAA Fisheries should bring together an independent, interdisciplinary panel of modelers and relevant empirical scientists that would meet periodically to pursue the modeling effort collaboratively in an iterative and adaptive manner with the long-term goal of developing tools to support informed, practical decision-making.

Species should be chosen on the basis of how extensively they have been studied, and the models should concentrate on populations (or subpopulations) in which individual animals are known and have been tracked for some time. The different species should be chosen to span an array of life-history patterns (such as feeding and breeding strategies). The objectives of the modeling exercises should be to speculate on how harassment or acoustic injury of individuals might affect populations and to identify gaps in data and understanding. The exercises should also explore links between IBMs and demographic analyses for the same population; each should be able to inform the other in important ways (see Caswell and John, 1992). Some candidate populations for such a study are the Puget Sound killer whales (Krahn et al., 2002), the North Atlantic right whales (*Eubalaena glacialis;* Waring et al., 2003), bottlenose dolphins in Sarasota Bay (Wells, 2003), the gray seals of Sable Island (Austin et al., 2004), and the northern elephant seals of Año Nuevo Island (LeBoeuf et al., 2000). All those have been studied extensively, and individual animals have been identified and resighted over multiple years. For most of the populations, the demographics are well defined; in some, the effects of major environmental stressors, such as an El Niño or the North Atlantic Oscillation, have been observed (Fujiwara and Caswell, 2001; Greene and Pershing, 2004). Such complex interdisciplinary modeling has been undertaken by the National Center for Ecological Analysis and Synthesis at the University of California, Santa Barbara.

4

Rational Management with Incomplete Data

The committee's task statement requires placing this scientific review within the context of management.

> Recognizing that the term "biologically significant" is increasingly used in resource management and conservation plans, this study will further describe the scientific basis of the term in the context of marine mammal conservation and management related to ocean noise.

As noted in this report, the full predictive model is at least a decade away from coming to fruition, and the management requirements involved in addressing concerns over ocean-noise effects on marine mammals are extremely pressing. Efforts are under way to address the long-term goal of producing the predictive model outlined here, but an interim plan is needed. One strategy is to implement a management regimen that uses available data, agreed upon management goals, and a conservative approach to the insufficiencies of the available data. The regimen should encourage data acquisition to reduce uncertainty. At the workshop the NOAA Fisheries Potential Biological Removal (PBR) model was discussed as such an example.

The three acts of Congress most relevant to regulating exposure of marine mammals to noise are the National Environmental Policy Act of 1969 (NEPA), the Marine Mammal Protection Act of 1972 (MMPA), and the Endangered Species Act of 1973 (ESA). The NEPA focuses on environmental analysis of "the relationship between local short-term uses of man's environment and the maintenance and enhancement of long-term productivity." The goal of the MMPA is to "replenish any species or population stock which has diminished below its optimum sustainable level," but its basic regulatory tool involves a prohibition on "taking" marine mammals, where *take* is defined as "to harass, hunt, capture or kill, or attempt to harass, hunt, capture or kill." Similarly, the ESA aims to "conserve endangered species and threatened species and the ecosystems upon which they depend," but it also relies on a prohibition of taking individual animals. The prohibition on taking marine mammals made sense when the dominant conservation problems involved directed hunting and animals incidentally killed by commercial fishing. It is much more difficult to relate harassment takes to population effects.

A number of the workshop panelists agreed that the concept of Potential Biological Removal (PBR) (Taylor et al., 2000) as developed by scientists at NOAA Fisheries, and the concept of the revised management procedure (Cooke, 1994) as developed by scientists associated with the International Whaling Commission, represented the best current approaches to management of human effects on marine mammals under conditions of inadequate data. This chapter reviews the PBR concept and suggests how harassment and other takes could be incorporated into it. The PBR concept is attractive because it is based on a small number of clearly defined and easily understood variables. The limits of acceptable population impact determine the allowable removals. Extensive modeling and sensitivity analysis confirmed that the selected parameter values ensured, with high probability, that the population impacts would be within the prescribed bounds. Anyone who feels that the allowed removals are set either too low or too high can present new data and interpretation in peer-reviewed publications that NOAA Fisheries uses in stock assessments and establishment of PBR.

FINDING: Development of a model, such as the PCAD model, to inform regulatory decisions is critical for a full understanding of the biological significance of anthropogenic noise on marine mammal populations, but a more immediate solution is necessary.

RECOMMENDATION 6: A practical process should be developed to help in assessing the likelihood that specific acoustic sources will have adverse effects on a marine mammal population by disrupting normal behavioral patterns. Such a

process should have characteristics similar to the Potential Biological Removal model, including

- Accuracy,
- Encouragement of precautionary management—that is more conservative (smaller removal allowed)—when there is greater uncertainty in the potential population effects of induced behavioral changes,
- Being readily understandable and defensible to the public, legal staff, and Congress,
- An iterative process that will improve risk estimates as data improve,
- An ability to evaluate cumulative impacts of multiple low-level effects, and
- Being constructed from a small number of parameters that are easy to estimate.

POTENTIAL BIOLOGICAL REMOVAL

The 1994 reauthorization of the MMPA introduced a new regime to determine when the number of animals killed or seriously injured by commercial fisheries poses a risk to marine mammal stocks. It involves estimating the number of animals that could be "removed" from a marine mammal stock without stopping the stock from reaching or maintaining its optimal sustainable population (16 U.S.C. 1362(3)20). The number is called the PBR. Under this regime, every fishing vessel is required to register with NOAA Fisheries. As long as the operators of the vessel register, accept an observer on board, report every marine mammal that they find killed or seriously injured, and comply with the requirements of regulations adopted under a take-reduction plan, all the requirements under the MMPA have been met. In effect, they are exempt from the prohibition on harassment.

For each marine mammal stock, the number of animals killed or seriously injured is compared with the PBR. If NOAA Fisheries learns of sources of mortality, such as a ship strike, the animals are added to the total, but there is no systematic effort to monitor nonfishing kills.[1] If the number

---

[1] From the Marine Mammal Commission's 2002 report to Congress: "The Commission also questioned the Service's decision to include data on fishery- and other human-related

of animals taken is above the PBR, the regimem calls for a take-reduction team to be formed and to determine ways to reduce the take. The take-reduction team is required to recommend management actions that will reduce the take to below the PBR within 6 months and to the zero-mortality goal within 5 years. A rule establishing 10% of the PBR as zero mortality was published in the July, 20 2004, *Federal Register*.

The calculation of the PBR provides an example of a model designed for management and decision-making. The criteria used for this model are these (Taylor et al., 2000):

- Input parameters are based on available data.
- Uncertainty is incorporated into the model. Managers must make decisions despite uncertainty, but decisions grow more conservative with greater uncertainty.
- There is a mechanism for demonstrating that decisions based on the model meet the MMPA management goals.

Before 1994, the MMPA prohibited any kills of marine mammals in stocks that were below an optimal sustainable population (OSP). The MMPA defines OSP on the basis of the theory of density-dependent population growth. The OSP is defined as the maximal net productivity level (MNPL), which is the population size that theoretically yields the greatest growth rate. The MMPA characterized populations that fell below the MNPL as depleted. During the first 20 years of the MMPA, however, it proved difficult to estimate the parameters required to determine when a population reached the critical point of depletion. Given that uncertainty and the draconian consequences of a "depleted" designation, few populations were designated as depleted, and depletion designations did not fare well in court.

The PBR model was developed in response to the difficulty in parameter estimation. The PBR model selected inputs on the basis of the

---

mortalities and serious injuries only when incidents could be confirmed. In the Commission's view, requiring confirmation runs counter to the precautionary principle built into the Marine Mammal Protection Act and would tend to result in underestimates. Similarly, the Commission took issue with conclusions in some assessment reports, particularly those for the Alaska region, that certain effects were not occurring because they had not been observed. The Commission cautioned that such conclusions of no-effect should be based, in part, on monitoring effort being made to detect such effects."

experience that the three parameters most easily estimated for most marine mammals were abundance, the uncertainty of abundance, and maximal growth rate. The PBR is calculated as follows:

$$PBR = 0.5 N_{min} R_{max} F_r$$

where N_{min} is the minimum population estimate, R_{max} is the maximal population growth rate, and F_r is a recovery factor ranging from 0.1 to 1.0. Qualitatively, it should be clear that the larger the population and the faster it is capable of growing, the more animals can be removed from the population without impeding its recovery. The equation for PBR was not derived from population modeling, however, but through modeling to evaluate its ability to meet, with a 95% probability, the following management goals based on the MMPA (Taylor et al., 2000):

- Healthy populations will remain above OSP numbers for the next 20 years.
- Recovering populations will reach OSP numbers after 100 years.
- Populations at high risk will not be delayed in reaching OSP numbers by more than 10% beyond the predicted time that is based on an absence of human-induced mortality.

Biologists at NOAA Fisheries tested various values for the input parameters to decide on the values most likely to meet management goals.

The PBR model incorporates two features that are desirable in a model to be used for management decisions (Taylor et al., 2000). It uses parameters that are readily available, and it is conservative when there is uncertainty. For example, the use of the minimal population estimate takes an immediately conservative approach while research to refine the population estimate is stimulated. That is particularly true when the take is near the PBR and the minimal population estimate leads to a PBR well below that calculated by using the mean population estimate. The validity of the PBR is based on how well the result meets explicit management objectives.

EXTENSION OF PBR

PBR should be extended in two ways. First, it needs to incorporate mortality outside the regulated fishing industries. Second, it needs to consider effects on populations that result from the summation of multiple sublethal impacts on individuals. Although the PBR regime was initially developed to regulate commercial fisheries, it cannot achieve the goals of the MMPA if activities other than fisheries contribute to mortality and these takes are not counted accurately and tallied with the fishery takes. For example, NOAA Fisheries has instituted a costly scheme of using professional monitors on vessels to count animals that are entangled in fishing gear, and fisheries are required to report deaths and serious injuries. In many fisheries, however, animals may be killed or injured in lost gear, and this is unlikely to be detected by monitoring on the fishing vessels (Laist, 1996). Similarly, animals immobilized in fishing gear may be taken by predators or may become disentangled after injury or death and not be counted. The regulations requiring reporting of lethal takes and serious injuries are limited to fisheries, so the accounting of takes in nonfishery activities is not as accurate.

The NOAA Fisheries stock assessments are improving their reporting of takes in such activities as vessel strikes, but without a reliable mechanism for monitoring and reporting it is nearly impossible to estimate the number of takes in a given activity. There may be additional uncounted lethal takes from a variety of sources, including exposure to intense noise.

The potential for such takes of Cuvier's beaked whales in association with naval sonar was reflected in the NOAA Fisheries 2002 stock assessment for Cuvier's beaked whales in the western North Atlantic. The assessment lists 46 fisheries-related beaked whale deaths from 1989 to 1998, 53 beaked whales stranded from 1992 to 2000, and 14 beaked whales stranded in the Bahamas in association with a naval sonar exercise. The assessment points out other associations between mass strandings of beaked whales and the presence of naval vessels (NMFS, 2002, pg. 50)

> Although a species-specific PBR cannot be determined, the permanent closure of the pelagic drift gillnet fishery has eliminated the principal known source of incidental fishery mortality. The total fishery mortality and serious injury for this group is less than 10% of the calculated PBR and, therefore can be considered to be insignificant and approaching zero mortality and serious injury rate. This is a strategic stock because of uncertainty regarding stock size and evidence of human induced mortality and serious injury associated with acoustic activities.

The stock assessment states that the stock is strategic because of acoustic activities, now that the fishery rate is low. This is a clear example of where the PBR mechanism cannot protect marine mammals unless NOAA

Fisheries develops a mechanism for accurate reporting of all sources of human-induced mortality.

FINDING: During the last decade, the PBR mechanism has proved to be a successful model to account for the cumulative effects of lethal takes and serious injuries in commercial fisheries. However, as currently implemented, the PBR mechanism cannot adequately protect marine mammals from all sources of human-induced mortality until all such mortality is included in a revised and expanded PBR regime.

> **RECOMMENDATION 7**: Improvements to PBR are needed to reflect total mortality losses and other cumulative impacts more accurately:
>
> - NOAA Fisheries should devise a revised PBR regime in which all sources of mortality and serious injury can be authorized, monitored, regulated, and reported in much the same manner as is currently done by commercial fisheries under Section 118 of the MMPA.
> - NOAA Fisheries should expand the PBR model to include injury and behavioral disturbance with appropriate weighting factors for severity of injury or significance of behavioral response (cf. NRC, 1994, p. 35).

The PBR is intended as a mechanism to trigger regulatory action when the cumulative effects of taking reach some threshold. It uses the number of individuals removed from the population as the unit for assessing cumulative effect. Individuals are taken when they are killed, but taking also includes serious injury, minor injury, and behavioral disturbance. Rather than the current practice of counting serious injury as equal to death and injury as equivalent to no effect, it would be appropriate to develop a severity score for each kind of take defined by the MMPA. A severity score estimates the proportional effect of a given take activity compared with that of a lethal take. A precise estimate of the proportion would require integration of behavioral effects into demographic models—one of the most challenging aspects of the PCAD model. However, it may be possible to set several categories of severity for injury and behavioral harassment. Two categories per order of magnitude would probably provide appropriate precision (for example, 1, 0.3, 0.1, 0.03, 0.01, 0.003).

The visible signs of injury listed by NOAA Fisheries[2] include injuries of obviously varied severity. They include

- Loss of or damage to an appendage, jaw, or eye; these injuries affect the long-term ability of an animal to swim, feed, or see.
- Entanglement in fishing gear; it may take days or weeks for an animal to free itself from a serious entanglement, which may also leave long-term injuries.
- Bleeding, laceration, swelling or hemorrhage; some of these may reflect a serious injury, but they often resolve in a few days with little long-term consequence.

To address Recommendation 7, NOAA Fisheries could convene an expert panel of veterinarians to assign injury severity scores for those and other symptoms. For example, it seems likely that the first category might score 0.3, the second category 0.1, and the third category 0.01. Although some of the animals with the symptoms may have more or less severe effects, as long as the severity score is at least as great as the effect on the average animal compared with being killed, the scoring should be conservative for use in the PBR. The research necessary to validate that would involve following the outcomes of injured animals for their ability to survive, grow, breed, and provide parental care.

Just as the cumulative effects of nonserious injuries cannot be ignored, so an analysis of cumulative effects must add the adverse effects of behavioral harassment. Behavioral harassment is likely to be both less severe and more common than injury. That makes it all the more important to evaluate the cumulative effects on a stock of all harassment takes in addition to injury and lethal takes. For example, the dominant model of effects of noise posits different zones of influence at different distances from the source (Figure 4-1).

Assigning a severity score to harassment would involve a process similar to that used for injury but would require experts in behavioral ecology instead of veterinary care. Assuming that harassment is not involved indirectly in causing injury or death (as may occur with effects of military sonar on beaked whales), the primary effects of harassment involve the loss of opportunities, time, and energy. If the proposed activity occurs at a criti-

---

[2]http://www.nmfs.noaa.gov/prot_res/PR2/Fisheries_Interactions/MMAP.htm

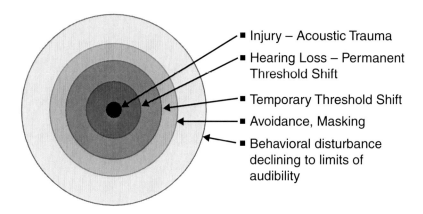

FIGURE 4-1. Close to an intense source, sound may be loud enough to cause death or serious injury. Somewhat farther away, an animal might have less serious injury, such as hearing loss. Temporary threshold shifts occur at greater distances. Animals may avoid exposures at even greater distances or they may not move from the area but still be affected through masking of important auditory cues from the environment. They may show just observable behavioral disturbance at distances comparable with the limit of audibility. The different distances for the different effects define different areas for each zone.
SOURCE: Modified from Richardson and Malme, 1993.

cal time or in a critical place when a specific activity must occur (for example, it disrupts a critical feeding trip of a phocid seal or disturbs a breeding site during a short season), the severity score will be higher. Thus, for a species for which the cost of a lost breeding season reflects the postponement to the next season and for an individual expected to have well in excess of 10 breeding seasons, the severity of loss of a breeding season might be set at 0.1; if the expectation is well in excess of 30 breeding seasons, the severity of loss of a breeding season might be set at 0.03. For activities that are expected to expose animals for shorter times during less critical periods, the time and energy lost may dominate interpretation of severity. One of the most pronounced behavioral responses of a marine mammal to noise involves the response of beluga whales to icebreakers in the Arctic. Beluga whales may respond to an icebreaker at many tens of kilometers (LGL and

Greeneridge, 1986; Cosens and Dueck, 1988; Finley et al., 1990). Their normal behavior is disrupted for several days, and they may have an increased metabolic rate as they swim away from an oncoming vessel. Other animals in other settings may show disruption of behavior for minutes to hours. In those cases, the severity score may be based on time lost and excess energy expended. Many species have seasonal changes in their behavioral ecology, with seasons lasting around 100 days, so a first approximation might divide the expected duration of disruption, in days, by 100. The result could be rounded to the next higher severity score. Thus, if an activity would be expected to disrupt an animal for less than 0.1 day (2.4 hr), the severity would be 0.1/100 = 0.001. If the disruption would be expected to last minutes, the severity might be set a .003/100 = 0.00003. As with the severity score for injury, an expert panel could be convened to establish severity scores for different kinds of behavioral disruption.

Severity scores can be used in the calculation of PBR by multiplying the number of animals affected by each severity (N) times the severity score (S) itself, and then tallying all of the N*S values. Table 4-1 illustrates the expectation that the higher the severity score, the fewer animals expected to be impacted, but in addition it illustrates how leaving out the cumulative effects of injury and harassment may underestimate cumulative impacts. In this hypothetical example, with an unrealistic assumed density of 1 animal/ 3.14 m^2, there is 1 lethal take, the equivalent of 1 lethal take in 10 injuries, and the equivalent of 1 lethal take in 100 cases of behavioral harassment. If PBR is to correctly tally cumulative impacts, it cannot completely ignore effects with severity of <1.

TABLE 4-1 Arbitrary Ranges and Severity Levels to Illustrate the Relation Between Severity of Effect and Numbers of Animals Affected (for most species, a two-dimensional approximation is appropriate)

Effect	Range (m)	Severity (S)	Relative Area (πr^2)	Number of Animals (N)	(N)*(S)
Death or serious injury	1	1	3	1	1
Injury (such as hearing loss)	10	0.01	314	100	1
Behavioral Disturbance	100	0.0001	31,416	10,000	1
TOTAL					3

DETERMINATION OF NONSIGNIFICANT IMPACT

The proposed modifications of the PBR model cannot be accomplished easily or quickly. The original PBR model was the result of many years of development and analysis. Prior sections of this report have emphasized the long time-line for acquiring the data and understanding necessary for a full implementation of the PCAD model. Compliance with the current regulatory interpretations of the NEPA, the MMPA, and the ESA is fraught with uncertainty regarding the use of sound sources in the marine environment and as the 2000 National Research Council report noted, regulations are more effective when they target critical disturbances.

The statement of task for this study was initially framed as identify biologically significant effects, but from a regulatory perspective it is more important right now to suggest a process for identifying activities that do *not* reach a de minimis standard for biological significance. Such activities would be exempt from the normal permitting process.

To assist regulatory agencies in meeting the requirements of the MMPA, a formalized, intelligent-decision system for risk assessment that uses current research expertise could offer the following advantages:

- It could provide a rapid and more simple authorization procedure, reducing the burden on applicants and regulators.
- It could provide a tally of each effect in a format that could account for cumulative effects.
- It could stimulate the generation of data required to make determinations in a format that makes the data readily available for the next applicant.
- It could improve decisions by improving available data.
- It could encourage others to report problems (such as, strandings) and to identify unexpected potential problems.
- It could set conditions for permits on the basis of location, time, and ecological conditions.
- It could maintain permanent records of every application.
- It could require applicants who apply and fail to meet a de minimis standard to obtain permits as under the current system.
- It could institute an adaptive system to improve data incrementally, and to reflect updates from annual reviews.

An Internet-based system, described in Figure 4-2 could assist producers of sound in the sea to determine whether proposed activities

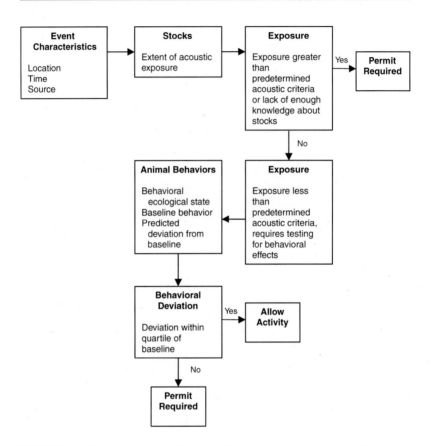

FIGURE 4-2. Diagram of a possible system for determination of whether behavioral changes cross a de minimis threshold.

require a permit or may be considered exempt from permitting. Essentially, such a process would allow regulators to establish de minimis standards that identify activities that have a low probability of causing changes in marine mammal behavior that would lead to significant population effects. This system would be populated initially with rules that, given our current state of knowledge, can best be attained through expert opinion. Although the model presented is based on animal exposure to sound, it is equally applicable to other types of activities affecting marine mammals.

In the initial stage of the process for applying for the de minimis exemption, for any kind of effect on marine mammals, the applicant would state the location and time of the proposed activity. The spatial scope of most effects is relatively easy to define. Sound travels so well in water that determining the scope of acoustic effects requires more information. For acoustic effects, the applicant would also state the acoustic characteristics of the proposed source: for example, source level, rise time, spectrum, directionality, and time course of operation.

Because most marine mammal populations are below their OSP, the system should be conservative in the face of uncertainty, that is, it should avoid the type of error that would lead to the loss of a valued resource (NRC, 1995). Such conservativeness might be reflected in a requirement for a specified level of knowledge about the distribution of animal populations, known as stocks for management purposes, within hearing range of the source. If enough is known about the stocks and their distribution, the system would move to the next stage; if not, it would reject the application for "no significant effect" determination unless the applicant could obtain and enter the required information.

The initial format of this part of the system would be based on a geographical information system (GIS). It could build on several continuing efforts to develop GIS systems that store information about the distribution and abundance of stocks (such as the Ocean Biogeographical Information System Spatial Ecological Analysis of Megavertebrate Populations, http://seamap.env.duke.edu) with geographical data on sound propagation. The common database described in Recommendation 3 could be used to populate this part of the system. The raw sighting data used by NOAA Fisheries for stock assessments would be a major component of the marine mammal element of the GIS for US waters. The acoustic information could be used to define how the sound would spread from the proposed site.

The initial stage in evaluating whether potential effects of a sound source cross the de minimis threshold would use the NOAA Fisheries acoustic criteria described in Chapter 3. For each species in the area, the exposure to sound from the planned sources is evaluated in terms of the criterion threshold for sound pressure level or energy level for the functional hearing group to which the species belongs. If the probability that individuals are exposed above the threshold level for acoustic effects is less than, for example, 0.001, the species would pass the proposed de minimis standard for direct acoustic exposure.

Animals experiencing exposures below the direct acoustic-effects threshold may still have behavioral reactions that could lead to population consequences. The next step is to determine the level of effect on life functions (Box 4-1).

**BOX 4-1
Considerations for Evaluating Marine Mammal
Disturbances by Specific Activities**

Determining biologically significant disturbance would necessarily evaluate a number of behaviors and their ecological contexts in regard to the proposed activities. Below are some behaviors that theoretically can be disrupted by noise, and some considerations in the determination of significance of the disruptions. The examples are illustrative only and should not be construed as a complete catalog of potentially biologically significant behavioral disruptions.

Migration. For migration, the standard might state that neither the path length nor the duration of migration could be increased into the upper quartile of the normal time or distance of migration. Fully one-fourth of the population exceed this value normally, so this is likely to be a conservative criterion. With enough data on time and length of migration, the applicant could then use response models or estimates of the scope of the effect to evaluate whether they meet the criterion. For example, if the effect of the activity extends for only a small duration of migration or a small part of the migratory path, such data alone might be sufficient. For migrating gray whales, in which case avoidance can be quantitatively related to a received level of sound, more-detailed analyses might be applied to a measure to account for the reduced uncertainty.

Feeding. For feeding behavior, the standard might be related to whether the disturbance will decrease energy reserves into the lower quartile of normal variation, as measured during a period appropriate for the proposed activity and season and the species affected. For example, female marine mammals can be divided into capital breeders, which postpone reproduction until they have stored enough energy to carry infants through to weaning, and income breeders, which continue to make foraging trips during lactation (Costa, 1993). Different periods would be integrated for the different classes and different energy measures, such as energy stores or reserves vs. daily energy balance.

The behavior of marine mammals varies by species, age-sex class, location, season, and time. The effect on life functions of a given change in behavior will also depend on those variables. The effect can be modulated

---

Breeding. Different standards for disruption of breeding behavior should be considered for females and males. The ability of a female to select a mate, breed, gestate, and give birth to a viable offspring is so essential to populations that there should be very low tolerance of disturbances that might affect these activities. The disruption of male reproductive behavior is probably less likely to have population effects than would disruption of female reproductive behavior, although disruption of male behavior should not reduce the pool of potential mates from which females can choose by more than 25%. This might be estimated from known changes in male call characteristics in response to noise, if the typical distribution of males and disturbance-caused movements of females are sufficiently known, the scope of disturbance could be estimated.

Nurturing and Parental Care. Very low thresholds should be considered for any disturbance that might separate a dependent infant from its caregivers. Examples include analyzing whether noise or disturbance responses might cause the infant and caregivers to separate too far to resume their activities. On longer time scales, the program could analyze whether the disturbance might reduce the nutrition from lactation to less than the lower quartile of normal. Both the duration of nursing bouts and the distribution of intervals between bouts may be important. It is possible that males in some species, such as Baird's beaked whales (*Berardius bairdii;* Kasuya et al., 1997), may be important for parental care and infant survival. Undisturbed social structure may be particularly important for infant survival. For example, bottlenose dolphin calves raised in large, more stable groups have higher survival than those raised in smaller, less stable groups (Wells, 1993).

Predator Avoidance. For behavioral changes that alter the response to predators, very low thresholds are recommended if there is the chance that the disruption will increase the vulnerability of an animal to predation. Many marine mammals depend on social defenses from predation (Mann et al., 2000).

by interannual ecological changes, such as El Niño or the North Atlantic Oscillation. Because the science is not mature enough for predictive modeling from behavior of individuals to population effects, a simple interim criterion based on normal variation of undisturbed behavior could be used. The baseline behavior against which behavioral changes are measured should be mapped onto the time and location of the proposed activity as closely as possible. Where other contexts, such as the phase of the interannual cycle, are known to affect behavior, they should be taken into account.

The de minimis criterion should be robust and conservative in the face of small samples and ignorance of shape of the distribution of baseline behavior. It should also be set at a level that meets management goals. A reasonable starting point would be a quartile level (upper or lower, as appropriate), but the value selected for this criterion should be tested with the same kinds of models used to evaluate the performance of the calculation of PBR (Taylor et al., 2000).

In all cases in which the proposed system yields a "no-significant-impact" determination and the applicant does not have to prepare a permit application, NOAA Fisheries should require the applicant to register the activity, monitor for effects, and report observed effects to the system to improve the knowledge base for future determinations. Approved stranding networks should enter all stranding data. The Internet-based system could be queried for any planned activities, and anyone could look for correlations between activities and strandings. After accumulating data for a few years, the database would allow epidemiological research that should be able to identify such problems as the effects of mid-range tactical sonar on beaked whales in less than the 35 years that it took to make this particular connection.

Experts and managers should meet annually, at least initially, to evaluate the performance of the system and to revise decision criteria on the basis of new information. Such a system, if applied to all activities, would provide rich opportunities for epidemiological analyses of the data to identify hot spots and linkages between human activities and marine mammal mortality or morbidity.

Any cases of lethal take or serious injury should be reported immediately and should be added to the take that is compared with the PBR. Any such take should disqualify the activity for the "no significant impact" determination and for regulation under the de minimis standard. Any applicant who provides false information to the system in an attempt to avoid permitting requirements should be disqualified from using the system and be subject to prosecution.

FINDING: Current knowledge is insufficient to predict which behavioral changes in response to anthropogenic sounds will result in significant population consequences for marine mammals. The PCAD model and proposed revisions to the PBR will take years to implement. In the interim, those who introduce sound into the marine environment and those who have responsibility for regulating takes resulting from such activities need a system whereby reasonable criteria can be set to determine which sounds will have a nonsignificant impact on marine mammal populations. Collectively, there are sufficient expert knowledge and extensive databases to establish such a system and to set the non-significant-impact criterion conservatively enough that there can be broad agreement on it.

RECOMMENDATION 8: An intelligent-decision system should be developed to determine a de minimis standard for allowing proposed sound-related activities. An expert-opinion panel should be constituted to populate the proposed system with as many decision points as current information and expert opinion allow. The system should be systematically reviewed and updated regularly.

References

Abbott, C.A., M.W. Berry, E.J. Comiskey, L.J. Gross and H.K. Luh. 1997. Parallel individual-based modeling of Everglades deer ecology. *IEEE Computational Science and Engineering* 4(4):60-72.

Alberts, S.C. and J. Altmann. 1995. Preparation and activation: determinants of age at reproductive maturity in male baboons. *Behavioral Ecology and Sociobiology* 36:397-406.

Alberts. S.C., R.M. Sapolsky and J. Altmann. 1992. Behavioral, endocrine, and immunological correlates of immigration by an aggressive male into a natural primate group. *Hormones and Behavior* 26:167-178.

Altmann, J. and P. Muruthi. 1988. Differences in daily life between semi-provisioned and wild-feeding baboons. *American Journal of Primatology* 15:213-222.

Altmann, J., G. Hausfater and S.A. Altmann. 1988. Determinants of reproductive success in savannah baboons *Papio cynocephalus*. Pp. 403-418 in *Reproductive Success*, T.H. Clutton-Brock, Ed. University of Chicago Press, Chicago, Illinois.

Altmann, J., D. Schoeller, S.A. Altmann, P. Muruthi and R.M. Sapolsky. 1993. Body size and fatness of free living baboons reflect food availability and activity levels. *American Journal of Primatology* 30:149-161.

Altmann, J. and S.C. Alberts. 2003. Intraspecific variability in fertility and offspring survival in a nonhuman primate: behavioral control of ecological and social sources. Pp. 140-169 in *Offspring: The Biodemography of Fertility and Family Behavior*, K.W. Wachter and R.A. Bulatao, Eds. National Academy Press, Washington, DC.

Andelman, S.J., S. Beissinger, J. Fitts Cochrane, L. Gerber, P. Gomez-Priego, C. Groves, J. Haufler, R. Holthausen, D. Lee, L. Maguire, B. Noon, K. Ralls and H. Regan. 2001. *Scientific standards for conducting viability assessments under the National Forest Management Act: report and recommendations of the NCEAS working Group*. National Center for Ecological Analysis and Synthesis, Santa Barbara, California.

Anderson, J.L. 1998. Embracing uncertainty. *Conservation Ecology* 2:2.

Austin, D., W.D. Bowen and J.I. McMillan. 2004. Intraspecific variation in movement patterns: Using quantitative analyses to characterize individual movement behaviour in a large marine predator. *Oikos* 105:15-30.

Barlow, J. and P.J. Clapham. 1997. A new birth-interval approach to estimating demographic parameters of humpback whales. *Ecology* 78:535-546.

Bauman, D.E. 2000. Regulation of nutrient partitioning during lactation: homeostasis and homeorhesis revisited. Pp. 311-327 in *Ruminant Physiology: Digestion, Metabolism and Growth and Reproduction*, P.J. Cronje, Ed. CAB Publishing, New York, New York..

Bekoff, M. and J.A. Byers. 1998. *Animal play: Evolutionary, comparative, and ecological perspectives.* Cambridge University Press, Cambridge, Massachusetts.

Block, B.A., D.P. Costa, G.W. Boehlert and R.E. Kochevar. 2003. Revealing pelagic habitat use: the tagging of Pacific pelagics program. *Oceanologica Acta* 25:255-266.

Boness, D.J. and W.D. Bowen. 1996. The evolution of maternal care in pinnipeds. *BioScience* 46:645-654.

Boyd, I. 2002. Energetics: consequences for fitness. Pp. 247-277 in *Marine Mammal Biology: an evolutionary approach*, A. Rus Hoelzel, Ed. Blackwell Science Ltd., Oxford, United Kingdom.

Brault, S. and H. Caswell. 1993. Pod-specific demography of killer whales (*Orcinus orca*). *Ecology* 74:1444-1454.

Brownell, R.L., Jr. and D.W. Weller. 2002. *Prolonged calving intervals in Western gray whales: nutritional stress and pregnancy.* Paper SC/54/BRG12 presented to the IWC Scientific Committee, April 2002. 13 pp. This document is available from the office of the International Whaling Commission, Cambridge, United Kingdom.

Brownell, R.L., Jr., T. Yamada, J.G. Mead and A. van Helden. 2004. *Mass strandings of Cuvier's beaked whales in Japan: U.S. Naval Acoustic Link?* Paper SC/56/E37 presented to the IWC Scientific Committee, July 2004. 10 pp. This document is available from the office of the International Whaling Commission, Cambridge, United Kingdom.

Bryant, P.J., C.M. Lafferty and S.K. Lafferty. 1984. Reoccupation of Laguna Guerrero Negro, Baja California, Mexico, by gray whales. Pp. 375-386 in *The Gray Whale Eschrichtius robustus*, M.L. Jones, et al., Eds. Academic Press, Orlando, Florida.

Buchanan, K.L. and A.R. Goldsmith. 2004. Noninvasive endocrine data for behavioral studies: the importance of validation. *Animal Behavior* 67:183-185.

Burgess, W.C. 2001. A general-purpose acoustic recording tag for marine wildlife. *Journal of the Acoustical Society of America* 110:2722-2723.

Caswell, H. 1989. *Matrix population models.* Sinauer Associates, Sunderland, Massachusetts.

Caswell, H. 2001. *Matrix population models.* Second edition. Sinauer Associates, Sunderland, Massachusetts.

Caswell, H. and M. Fujiwara. 2004. Beyond survival estimation: mark-recapture, matrix population models, and population dynamics. *Animal Biodiversity and Conservation* (in press).

Caswell, H. and A.M. John. 1992. From the individual to the population in demographic models. Pp. 36-61 in *Individual-based Models and Approaches in Ecology*, D. Deangelis and L. Gross, Eds. Chapman and Hill, New York, New York.

Caswell, H., S. Brault, A. Read and T. Smith. 1998. Harbor porpoise and fisheries: an uncertainty analysis of incidental mortality. *Ecological Applications* 8:1226-1238.

Cavieres, M.F., J. Jaeger and W. Porter. 2002. Developmental toxicity of a commercial herbicide mixture in mice: 1. Effects on embryo implantation and litter size. *Environmental Health Perspectives* 110:1081-1085.

Cavigelli, S.A. and M.K. McClintock. 2003. Fear of novelty in infant rats predicts adult corticosterone dynamics and an early death. *Proceedings of the National Academy of Sciences* 100:16131-16136.

Clapham, P.J., S.B. Young and R.L. Brownell Jr. 1999. Baleen whales: conservation issues and the status of the most endangered populations. *Mammal Review* 29:35-60.

Cooke, J.G. 1994. The management of whaling. *Aquatic Mammalogy* 20:129-135.

Cosens, S.E. and L.P. Dueck. 1988. Responses of migrating narwhal and beluga to icebreaker traffic at the Admiralty Inlet ice-edge, N.W.T. in 1986. Pp. 39-54 in *Port and Ocean Engineering Under Arctic Conditions, Volume II*, W.M. Sackinger, et al., Eds. Geophysics Institute, University of Alaska, Fairbanks, 111pp.

Costa, D.P., B.J. Le Boeuf, A.C. Huntley and C.L. Ortiz. 1986. The energetics of lactation in the northern elephant seal, *Mirounga angustirostris*. *Journal of Zoology Series A* 209:21-34.

Costa, D.P. 1991. Reproductive and foraging energetics of pinnipeds: implications for life-history patterns. Pp. 300-344 in *Behavior of Pinnipeds*, D. Renouf, Ed. Chapman and Hall, London, United Kingdom.

Costa, D.P. 1993. The relationship between reproductive and foraging energetics in the evolution of the Pinnipedia. *Symposia of the Zoological Society of London* 66:293-314.

Costa, D.P., D.E. Crocker, J. Gedamke, P.M. Webb, D.S. Houser, S.B. Blackwell, D. Waples, S.A. Hayes and B.J. Le Boeuf. 2003. The effect of a low-frequency sound source (acoustic thermometry of the ocean climate) on the diving behavior of juvenile northern elephant seals, *Mirounga angustirostris*. *Journal of the Acoustical Society of America* 113(2):1155-1165.

Cox, T.M., A.J. Read, A. Solow and N. Trengenza. 2001. Will harbour porpoises (*Phocoena phocoena*) habituate to pingers? *Journal of Cetacean Research and Management* 3(1):81-86.

Creel, S., J.E. Fox, A. Hardy, J. Sands, B. Garrott and R.O. Peterson. 2002. Snowmobile activity and glucocorticoid stress responses in wolves and elk. *Conservation Biology* 16(3):809-814.

Croll, D.A., C.W. Clark, J. Calambokidis, W.T. Ellison and B.R. Tershy. 2001. Effect of anthropogenic low frequency noise on the foraging ecology of *Balaenoptera* whales. *Animal Conservation* 4:13-27.

Crouse, D.T., L.B. Crowder and H. Caswell. 1987. A stage-based population model for loggerhead sea turtles and implications for conservation. *Ecology* 68:1412-1423.

Crum, L. A. and Y. Mao. 1996. Acoustically enhanced bubble growth at low frequencies and its implications for human diver and marine mammal safety. *Journal of the Acoustical Society of America* 99:2898-2907.

D'Amico, A. Ed. 1998. *Summary record SACLANTCEN Bioacoustics Panel*, La Spezia, Italy, 15-17 June, 1998. SACLANTCEN M-133, SACLANT Undersea Research Centre, San Diego, California, 128 pp. [Online] Available at: http://enterprise.spawar.navy.mil/nepa/whales/bioacoust.cfm [September 22, 2004].

Deecke, V.B., P.J.B. Slater and J.K.B. Ford. 2002. Selective habituation shapes acoustic predator recognition in harbour seals. *Nature* 420:171-173.

Doroff, A.M., J.A. Estes, T.M. Tinker, D.M. Burn and T.J. Evans. 2003. Sea otter population declines in the Aleutian archipelago. *Journal of Mammalogy* 84:55-64.

Evans, D.L. and G.R. England. 2001. *Joint interim report: Bahamas marine mammal stranding event of 15-16 March 2000*. US Department of Commerce and US Navy, Government Printing Press, Washington, DC, 59 pp. [Online]. Available at: http://www.nmfs.noaa.gov/prot_res/PR2/Health_and_Stranding_Response_Program/Interim_Bahamas_Report.pdf [May 25, 2004].

Evans, G.W., S. Hygge and M. Bullinger. 1995. Chronic noise and psychological stress. *Psychological Science* 6:333-338.

Evans, G.W., P. Lercher, M. Meis, H. Ising and W.W. Kofler. 2001. Community noise exposure and stress in children. *Journal of the Acoustical Society of America* 109:1023-1027.

Finley, K.J., G.W. Miller, R.A. Davis and C.R. Greene. 1990. Reactions of belugas, (*Delphinapterus leucas*) and narwhals (*Monodon monoceros*) to ice-breaking ships in the Canadian high arctic. *Canadian Bulletin of Fisheries and Aquatic Science* 224:97-117.

Finneran, J.J. 2003. Whole-lung resonance in a bottlenose dolphin (*Tursiops truncatus*) and white whale (*Delphinapterus leucas*). *Journal of the Acoustical Society of America* 114:529-535.

Finneran, J.J. and C.E. Schlundt. 2004. *Effects of intense pure tones on the behavior of trained odontocetes*. Report of the Space and Naval Warfare Systems Center, San Diego, California, 15pp.

Finneran, J.J., C.E. Schlundt, D.A. Carder, J. Clark, J.A. Young, J.B. Gaspin and S.H. Ridgway. 2000. Auditory and behavioral responses of bottlenose dolphins (*Tursiops truncatus*) and a beluga whale (*Delphinapterus leucas*) to impulsive sounds resembling distant signatures of underwater explosions. *Journal of the Acoustical Society of America* 108(1):417-431.

Finneran, J.J., C.E. Schlundt, R. Dear, D.A. Carder and S.H. Ridgway. 2002. Temporary shift in masked hearing thresholds in odontocetes after exposure to single underwater impulses from a seismic watergun. *Journal of the Acoustical Society of America* 111(6):2929-2940.

Fish, J.F. and J.S. Vania. 1971. Killer whale (*Orcinus orca*) sounds repel white whales (*Delphinapterus leucas*). *Fishery Bulletin* 69:531-535.

Frankel, A.S., W.T. Ellison and J. Buchanan. 2002. Application of the Acoustic Integration Model (AIM) to predict and minimize environmental impacts. *Oceans '02 MTS/IEEE* 3:1438-1443.

Fristrup, K.M., L.T. Hatch and C.W. Clark. 2003. Variation in humpback whale (*Megaptera novaeangliae*) song length in relation to low-frequency sound broadcasts. *Journal of the Acoustical Society of America* 113(6):3411-3424.

Frost, K.J., L.F. Lowry and R.R. Nelson. 1984. *Belukha whale studies in Bristol Bay, Alaska*. Workshop on biological interactions among marine mammals and commercial fisheries in the southeastern Bering Sea, University of Alaska, Anchorage.

Fujiwara, M. 2002. Mark-recapture statistics and demographic analysis. Ph.D. Dissertation, Massachusetts Institute of Technology (MIT) and Woods Hole Oceanographic Institution Joint Program (WHOI) in Oceanography/Applied Ocean Science and Engineering, MIT, Cambridge and WHOI, Woods Hole, Massachusetts.

Fujiwara, M. and H. Caswell. 2001. Demography of the endangered North Atlantic right whale. *Nature* 414:537-541.

Fujiwara, M. and H. Caswell. 2002a. Estimating population projection matrices from multi-stage mark-recapture data. *Ecology* 83:3257-3265

Fujiwara, M. and H. Caswell. 2002b. A general approach to temporary emigration in mark-recapture analysis. *Ecology* 83:3266-3275.

Geraci, J.R. and V.J. Lounsbury. 1993. *Marine mammals ashore: A field guide for Strandings.* Texas A&M University Sea Grant College Program, Galveston, Texas.

Goodwin, P. and G. Wright. 1991. *Decision analysis for management judgment.* John Wiley & Sons, Chichester, United Kingdom.

Goold, J.C. 1996. Acoustic assessment of populations of common dolphin *Delphinus delphis* in conjunction with seismic surveying. *Journal of the Marine Biology Association, United Kingdom* 76:811-820.

Goymann, W. and J.C. Wingfield. 2004. Allostatic load, social status and stress hormones: The costs of social status matter. *Animal Behavior* 67:591-602.

Greene, C.H. and A.J. Pershing. 2004. Climate and the conservation biology of North Atlantic right whales: the right whale at the wrong time? *Frontiers in Ecology and the Environment* 2:29-34.

Grimm, V. 1999. Ten years of individual-based modeling in ecology: what have we learned and what could we learn in the future? *Ecological Modeling* 115:129-148.

Gross, K., B.A. Craig and W.D. Hutchinson. 2002. Bayesian estimation of a demographic matrix model from stage-frequency data. *Ecology* 83:3285-3298.

Harwood, J. 2000. Risk assessment and decision analysis in conservation. *Biological Conservation* 95:219-226.

Harwood, J. 2002. Mitigating the effects of acoustic disturbance in the oceans. *Aquatic Conservation: Marine and Freshwater Ecosystems* 12:485-8.

Heide-Jorgensen, M.P., T. Harkonen and P. Aberg. 1992. Long-term effects of epizootic in harbor seals in the Kattegat-Skagerrak and adjacent areas. *Ambio* 21:511-516.

Hermann, A.J., S. Hinckley, B.A. Megrey and J.M. Napp. 2001. Applied and theoretical considerations for constructing spatially explicit Individual-Based Models of marine fish early life history which include multiple trophic levels. *ICES Journal of Marine Science* 58:1030-1041.

Hill, R.D. 1986. Microcomputer monitor and blood sampler for free-diving Weddell seals. *Journal of Applied Physiology* 61:1570-1576.

Holling, C.S., Ed. 1978. *Adaptive Environmental Assessment and Management.* Wiley, Chichester, United Kingdom.

Houser, D.S., R. Howard and S. Ridgway. 2001. Can diving-induced tissue nitrogen supersaturation increase the chance of acoustically driven bubble growth in marine mammals? *Journal of Theoretical Biology* 213:183-195.

Hunt, K.E., A.W. Trites and S.K. Wasser. 2004. Validation of a fecal glucocorticoid assay for Steller sea lions (*Eumetopias jubatus*). *Physiology and Behavior* 80:595-601.

Hunter, C.M., H. Moller and D. Fletcher. 2000. Parameter uncertainty and elasticity analyses of a population model: setting research priorities for shearwaters. *Ecological Modelling* 134:299-323.

Ising, H. and B. Kruppa. 2004. Health effects caused by noise: Evidence in the literature from the past 25 years. *Noise & Health* 6:5-13.

Jepson, P.D., M. Arbelo, R. Deaville, I.A.P. Patterson, P. Castro, J.R. Baker, E. Degollada, H.M. Ross, P. Herráez, A.M. Pocknell, F. Rodríguez, F.E. Howie, A. Espinosa, R.J. Reid, J.R. Jaber, V. Martin, A.A. Cunningham and A. Fernández. 2003. Was sonar responsible for a spate of whale deaths after an Atlantic military exercise? *Nature* 425:575-576.

Johnson, M. and P.L. Tyack. 2003. A digital acoustic recording tag for measuring the response of wild marine mammals to sound. *IEEE Journal of Oceanic Engineering* 28:3-12.

Johnson, S.R. 2002. Marine mammal mitigation and monitoring program for the 2001 Odoptu 3-D seismic survey, Sakhalin Island, Russia. Paper SC/02/WGW19 presented to the IWC Scientific Committee, April 2002. 49 pp. This document is available from the Office of the International Whaling Commission, Cambridge, United Kingdom.

Kastak, D. and R.J. Schusterman. 1996. Sensitization and habituation to underwater sound by captive pinnipeds. *Journal of the Acoustical Society of America* 99(4):2577.

Kastak, D., R.J. Schusterman, B.L. Southall and C.J. Reichmuth. 1999. Underwater temporary threshold shift induced by octave-band noise in three species of pinniped. *Journal of the Acoustical Society of America* 106:1142-1148.

Kastelein, R.A., D. de Haan, A.D. Goodson, C. Staal and N. Vaughan. 1997. The effects of various sounds on a harbour porpoise (*Phocoena phocoena*). Pp. 367-383 in *The Biology of the Harbour Porpoise*, A.J. Read, P.R. Wiepkema and P.E. Nachtigall, Eds. De Spil Publishers, Woerden, Netherlands.

Kasuya, T., R.L. Brownell, Jr. and K.C. Balcomb, III. 1997. Life history of Baird's beaked whales off the Pacific coast of Japan. *Report of the International Whaling Commission* 47:969-979.

Ketten, D.R. 1994. Functional analyses of whale ears: Adaptations for underwater hearing. *IEEE Proceedings in Underwater Acoustics* 1:264-270.

Kokko, H., J. Lindström and E. Ranta. 1997. Risk analysis of hunting of seal populations in the Baltic. *Conservation Biology* 11:917-927.

Krahn, M.M., P.R. Wade, S.T. Kalinowski, M.E. Dahlheim, B.L. Taylor, M.B. Hanson, G.M. Ylitalo, R.P. Angliss, J.E. Stein and R.S. Waples. 2002. *Status review of Southern Resident killer whales (*Orcinus orca*) under the Endangered Species Act.* U.S. Department of Commerce, NOAA Technical Memorandum NMFS-NWFSC-54, 133 pp. National Oceanic and Atmospheric Administration, Silver Spring, Maryland.

Kuenzel, W.J., M.M. Beck and R. Teruyama. 1999. Neural sites and pathways regulating food intake in birds: a comparative analysis to mammalian systems. *Journal of Experimental Zoology* 283:348-364.

Laist, D.W. 1996. Impacts of marine debris: Entanglement of marine life in marine debris including a comprehensive list of species with entanglement and ingestions records. Pp. 99-139 in *Marine Debris: Sources, Impacts, and Solutions*. J.M. Coe and D.R. Rogers, Eds. Springer-Verlag, New York, New York.

Laist, D.W., A.R. Knowlton, J.G. Mead, A.S. Collet and M. Podesta. 2001. Collisions between ships and whales. *Marine Mammal Science* 17:35-75.

Lalas, C. and C.J.A. Bradshaw. 2003. Expectations for population growth at new breeding locations for the vulnerable New Zealand sea lion (*Phocarctos hookeri*) using a simulation model. *Biological Conservation* 114:67-78.

REFERENCES

Larson, S., C.J. Casson and S. Wasser. 2003. Noninvasive reproductive steroid hormone estimates from fecal samples of captive female sea otters (*Enhydra lutris*). *General and Comparative Endocrinology* 134:18-25.

Lavigne, D.M., W. Barchard, S. Innis and N.A. Øritsland. 1982. Pinniped bioenergetics. Pp. 191-235 in *Mammals of the Seas*, Volume V. Food and Agriculture Organization, Rome, Italy.

Le Boeuf, B.J., D.E. Crocker, D.P. Costa, S.B. Blackwell, P.M. Webb and D.S. Houser. 2000. Foraging ecology of northern elephant seals. *Ecological Monographs* 70:353-382.

LGL and Greeneridge. 1986. *Reactions of beluga whales and narwhals to ship traffic and ice-breaking along ice edges in the eastern Canadian High Arctic 1982-1984*. Environmental Studies 37. Indian and Northern Affairs Canada, Ottawa, Canada, 301 pp.

Lockyer, C. 1981. Growth and energy projects of large baleen whales from the southern hemisphere. Pp. 379-488 in *Mammals of the Seas*, Volume III. Food and Agriculture Organization, Rome, Italy.

Ludwig, D. 1996. Uncertainty and the assessment of extinction probabilities. *Ecological Applications* 6:1067-1076.

Malme, C.I., P.R. Miles, C.W. Clark, P. Tyack and J.E. Bird. 1983. *Investigations on the potential effects of underwater noise from petroleum industry activities on migrating gray whale behavior*. Report No. 5366 submitted to the Minerals Management Service, U.S. Department of the Interior, NTIS PB86-174174, Government Printing Press, Washington, DC.

Malme, C.I., P.R. Miles, C.W. Clark, P. Tyack and J.E. Bird. 1984. *Investigations on the potential effects of underwater noise from petroleum industry activities on migrating gray whale behavior. Phase II: January 1984 migration.* Report No. 5586 submitted to the Minerals Management Service, U.S. Department of the Interior, NTIS PB86-218377. Government Printing Press, Washington, DC.

Mann, J., R. Connor, P.L. Tyack and H. Whitehead, Eds. 2000. *Cetacean Societies: field studies of whales and dolphins*. University of Chicago Press, Chicago, Illinois.

Marine Mammal Commission. 2004. *Beaked Whale Technical Workshop*, Baltimore, Maryland, 13-16 April, 2004. Marine Mammal Commission, Bethesda, Maryland. [Online] Available at: http://www.mmc.gov/sound/beakedwhalewrkshp/beakedwhalewrkshp.html [October 28, 2004].

McEwen, B.S. 2000. Allostasis and allostatic load: Implications for neuropsychopharmacology. *Neuropsychopharmacology* 22:108-124.

McEwen B.S. and E. Stellar. 1993. Stress and the individual: Mechanisms leading to disease. *Archives of Internal Medicine* 153:2093-2101.

McEwen, B.S. and J.C. Wingfield. 2003. The concept of allostasis in biology and biomedicine. *Hormones and Behavior* 43:2-15.

Mertz, D.B. 1971. The mathematical demography of the California condor population. *American Naturalist* 105:437-453.

Meyer, M. and J. Booker. 1991. *Eliciting and Analyzing Expert Judgment, a Practical Guide*. Academic Press, London, United Kingdom.

Miller, P.J.O., N. Biasson, A. Samuels and P.L. Tyack. 2000. Whale songs lengthen in response to sonar. *Nature* 405:903.

Morgan, M.G. and M. Henrion. 1990. *Uncertainty. A Guide to Dealing with Uncertainty in Qualitative Risk and Policy Analysis.* Cambridge University Press, Cambridge, United Kingdom.

Morton, A.B. and H.K. Symonds. 2002. Displacement of *Orcinus orca* (L.) by high amplitude sound in British Columbia, Canada. *ICES Journal of Marine Science* 59:71-80.

Moulton, V.D., W.J. Richardson, M.T. Williams and S.B. Blackwell. 2003. Ringed seal densities and noise near an icebound artificial island with construction and drilling. *Acoustics Research Letters Online-ARLO* 4:112-117. [Online] Available at: http://scitation.aip.org/getabs/servlet/GetabsServlet?prog=normal&id=ARLOFJ000004000004000112000001&idtype=cvips&gifs=yes [September 22, 2004].

Munk, W.H., R.C. Spindel, A. Baggeroer and T.G. Birdsall. 1994. The Heard Island Feasibility Test. *Journal of the Acoustical Society of America* 96(4):2330-2342.

Muruthi, P., J. Altmann and S. Altmann. 1991. Resource base, parity, and reproductive condition affect females, feeding time and nutrient intake within and between groups of a baboon population. *Oecologica* 87:467-472.

Nachtigall, P.E., J.L. Pawloski and W.W.L. Au. 2003. Temporary threshold shifts and recovery following noise exposure in the Atlantic bottlenosed dolphin (*Tursiops truncatus*). *Journal of the Acoustical Society of America* 113:3425-3429

Nachtigall, P.E., A.Ya. Supin, J.L. Pawloski and W.W.L. Au. 2004. Temporary threshold shifts after noise exposure in the bottlenose dolphin (*Tursiops truncatus*) measured using evoked auditory potentials. *Marine Mammal Science* 20:673-687.

National Marine Fisheries Service. 2002. Cuvier's beaked whale (*Ziphius cavirostris*): Western North Atlantic Stock Pp. 46-51 in *U.S. Atlantic and Gulf of Mexico Marine Mammal Stock Assessments - 2002.* NOAA Technical Memorandum NMFS-NE-169. Department of Commerce, National Oceanic and Atmospheric Administration, Silver Spring, Maryland.

National Oceanic and Atmospheric Administration (NOAA). 2002. *Report of the Workshop on Acoustic Resonance as a Source of Tissue Trauma in Cetaceans,* Silver Spring, Maryland, 24-25 April, 2002. U.S. Department of Commerce, National Oceanic and Atmospheric Administration, National Marine Fisheries Service, Silver Spring, Maryland. [Online] Available at http://www.nmfs.noaa.gov/prot_res/readingrm/MMSURTASS/Res_Wkshp_Rpt_Fin.PDF. 19pp.

National Research Council (NRC). 1983. *Risk Assessment in the Federal Government: Managing the Process.* National Academy Press, Washington, DC.

National Research Council (NRC). 1993. *Issues in Risk Assessment.* National Academy Press, Washington, DC.

National Research Council (NRC). 1994. *Low-Frequency Sound and Marine Mammals: Current Knowledge and Research Needs.* National Academy Press, Washington, DC.

National Research Council (NRC). 1995. *Science and the Endangered Species Act.* National Academy Press, Washington, DC.

National Research Council (NRC). 2000. *Marine Mammals and Low-Frequency Sound.* National Academy Press, Washington, DC.

National Research Council (NRC). 2003a. *The Decline of the Steller Sea Lion in Alaskan Waters: Untangling Food Webs and Fishing Nets.* The National Academies Press, Washington, DC.

National Research Council (NRC). 2003b. *Ocean Noise and Marine Mammals*. The National Academies Press, Washington, DC.

Nichols, J.D., G.L. Hensler and P.W. Sykes, Jr. 1980. Demography of the Everglade kite: implications for population management. *Ecological Modeling* 9:215-232.

Nichols, J.D., J.R. Sauer, K.H. Pollock and J.B. Hestbeck. 1992. Estimating transition probabilities for stage-based population projection matrices using capture-recapture data. *Ecology* 73:306-312.

Norris, J.C., W.E. Evans and S. Rankin. 2000. An acoustic survey of cetaceans in the northern Gulf of Mexico. Pp. 173-216 in *Cetaceans, sea turtles and seabirds in the northern Gulf of Mexico: distribution, abundance and habitat associations*, R.W. Davis, W.E. Evans and B. Würsig, Eds. Final report, Technical report. USGS/BRD/CR-1999-0005/OCS Study MMS 2000-003. U.S. Department of the Interior, Minerals Management Service, Washington, DC, 346 pp.

Nowacek, D.P., M.P. Johnson and P.L. Tyack. 2003. North Atlantic right whales (*Eubalaena glacialis*) ignore ships but respond to alerting stimuli. *Proceedings of the Royal Society of London* 271:227-231.

Piantadosi, C.A. and E.D. Thalmann. 2004. *Pathology: Whales, sonar and decompression sickness (reply)*. Nature 428(6984):U1.

Pitcher, K.W. 1990. Major decline in number of harbor seals, *Phoca vitulina richardsi*, on Tugidak Island, Gulf of Alaska. *Marine Mammal Science* 6:121-134.

Porter, W.P., J.W. Jaeger and I.H. Carlson. 1999. Endocrine, immune, and behavioral effects of aldicarb (cabamate), atrazine (triazine) and nitrate (fertilizer) mixtures at groundwater concentrations. *Toxicology and Industrial Health* 15:133-150.

Ralls, K. and B.L. Taylor. 2000. Better policy and management decisions through explicit analysis of uncertainty: new approaches from marine conservation biology. *Conservation Biology* 14:1240-1242.

Rankin, S. 1999. *The potential effects of sounds from seismic exploration on the distribution of cetaceans in the northern Gulf of Mexico*. Report by the Department of Oceanography. Texas A&M University, College Station, 65 pp.

Richardson, W.J. and C.I. Malme. 1993. Man-made noise and behavioral responses. Pp. 631-700 in *The Bowhead Whale*, J.J. Burns, et al., Eds. Special Publication No. 2, Society for Marine Mammalogy, Lawrence, Kansas.

Ridgway, S. and R. Howard. 1979. Dolphin lung collapse and intramuscular circulation during free diving: evidence from nitrogen washout. *Science* 166:1182-1183.

Rolland, R.M., K.E. Hunt, S.D. Kraus and S.K. Wasser. 2004. Non-invasive measurement of reproductive hormones in North Atlantic right whales (*Eubalaena glacialis*). *Canadian Journal of Zoology* (In review).

Romero, L.M. 2004. Physiological stress in ecology: lessons from biological research. *Trends in Ecology and Evolution* 19:217-255.

Romero, L.M. and M. Wikelski. 2001. Corticosterone levels predict survival probabilities of Galapagos marine iguanas during El Niño events. *Proceedings of the National Academy of Sciences* 98(13):7366-7370.

Romero, L.M. and M. Wikelski. 2002. Severe effects of low-level oil contamination on wildlife predicted by the corticosterone-stress response: preliminary data and a research agenda. *Spill Science and Technology Bulletin* 7:309-313.

Sands, J. and S. Creel. 2004. Social dominance, aggression, and faecal glucocorticoid levels in a wild population of wolves, *Canis lupus. Animal Behaviour* 67:387-396.

Sapolsky, R.M. 1994. Individual differences and the stress response. *Seminars in the Neurosciences* 6:261-269.

Sapolsky, R.M., S.C. Alberts and J. Altmann. 1997. Hypercortisolism associated with social subordinance or social isolation among wild baboons. *Archives of General Psychiatry* 54:1137-1143.

Scientific Committee on Antarctic Research (SCAR). 2004. *SCAR report on marine acoustic technology and the Antarctic environment.* Information Paper IP 078. [Online] Available at: http://www.cep.aq/MediaLibrary/asset/MediaItems/ ml_381324289814815_27IP078E.doc [October 28, 2004].

Schleidt, W. 1961a. Reaktionen von Truthühnern auf fliegende Raubvögeln und Versuche zur Analyse ihrer AAM's. *Z. Tierpsychol* 18:534-560.

Schleidt, W. 1961b. Über die Auslösung der Flucht vor Raubvögeln bei Truthühnern. *Naturwissenschaften* 48:141-142.

Schlundt, C.E., J.J. Finneran, D.A. Carder and S.H. Ridgway. 2000. Temporary shift in masked hearing thresholds of bottlenose dolphins, *Tursiops truncatus*, and white whales, *Delphinapterus leucas*, after exposure to intense tones. *Journal of the Acoustical Society of America* 107:3496-3508.

Seeman, T.E., B.S. McEwen, J.W. Rowe and B.H. Singer. 2001. Allostatic load as a marker of cumulative biological risk: MacArthur studies of successful aging. *Proceedings of the National Academy of Sciences* 98(8):4770-4775.

Silk, J.B., S.C. Alberts and J. Altmann. 2003. Social bonds of female baboons enhance infant survival. *Science* 302:1231-1234.

Slooten, E., D. Fletcher and B. Taylor. 2000. Accounting for uncertainty in risk assessment: Case study of Hector's dolphin mortality due to gillnet entanglement. *Conservation Biology* 14:1264-1270.

Starfield, A.M. and A.L. Bleloch. 1991. *Building Models for Conservation and Wildlife Management* (second edition). Interactive Book Company, Edina, Minnesota.

Stevens, D.W. and J.R. Krebs. 1987. *Foraging Theory.* Princeton University Press, Princeton, New Jersey.

Stone, C.J. 2001. Marine mammal observations during seismic surveys in 1999. *JNCC Report.* No. 316 JNCC, Peterborough, United Kingdom.

Stone, C.J. 2003. *The effects of seismic activity on marine mammals in UK waters, 1998-2000.* JNCC Report No. 323. Report prepared for the Joint Nature Conservation Committee, Peterborough, United Kingdom.

Swartz, R.L. and R.J. Hofman. 1991. *Marine Mammal and Habitat Monitoring: Requirements, Principles, Needs, and Approaches.* Report prepared for Marine Mammal Commission. NTIS PB91-215046. National Technical Information Service, Springfield, Virginia.

Taylor, B.L., P.R. Wade, R.A. Stehn and J.F. Cochrane. 1996. A Bayesian approach to classification criteria for Spectacled Eiders. *Ecological Applications* 6:1077-1089.

Taylor, B.L., P.R. Wade, D.P. DeMaster and J. Barlow. 2000. Incorporating uncertainty into management models for marine mammals. *Conservation Biology* 14:1243-1252.

Taylor, B.L., J. Barlow, R. Pitman, L. Balance, T. Klinger, D. DeMaster, J. Hildebrand, J. Urban, D. Palacios and J.G. Mead. 2004. *A call for research to assess risk of acoustic impact on beaked whale populations.* Paper SC/56/E36 presented to the IWC Scientific Committee, July 2004. 4 pp. This document is available from the office of the International Whaling Commission, Cambridge, United Kingdom.

Trillmich, F. and T. Dellinger. 1991. The effects of El Niño on Galapagos pinnipeds. Pp. 66-74 in: *Pinnipeds and el Nino: Responses to Environmental Stress,* F. Trillmich and K. Ono, Eds. Springer-Verlag, Heidelberg, Germany.

Trillmich, F. and K.A. Ono, Eds. 1991. *Pinnipeds and El Niño: Responses to Environmental Stress.* Springer-Verlag, Berlin, Germany.

Tukey, J.W. 1977. *Exploratory Data Analysis.* Addison-Wesley, Reading, Massachusetts.

Tyack, P. 1998. Acoustic communication under the sea. Pp. 163-220 in *Animal Acoustic Communication: Recent Technical Advances,* S.L. Hoop, M.J. Owren, and C.S. Evans, Eds. Springer-Verlag, Heidelberg, Germany.

Tyack, P.L. and C.W. Clark. 1998. *Quick-look report: Playback of low-frequency sound to gray whales migrating past the central California coast.* Woods Hole Oceanographic Institution, Woods Hole, Massachusetts.

Tyack, P.L. and C.W. Clark. 2000. Communication and acoustical behavior in dolphins and whales. Pp. 156-224 in *Hearing by Whales and Dolphins. Springer Handbook of Auditory Research,* W.W.L. Au, A.N. Popper and R.R. Fay, Eds. Springer-Verlag, New York, New York.

Tyack, P., J. Gordon and D. Thompson. 2004. Controlled Exposure Experiments to Determine the Effects of Noise on Large Marine Mammals. *Marine Technology Society Journal* 37(4):41-53.

U.S. Commission on Ocean Policy. 2004. *An Ocean Blueprint for the 21st Century: Final Report of the U.S. Commission on Ocean Policy.* U.S. Commission on Ocean Policy, Washington, D.C. [Online] Available at: http://www.oceancommission.gov/documents/welcome.html [November 10, 2004].

Wade, P.R. 2000. Bayesian methods in conservation biology. *Conservation Biology* 14:1308-1316.

Ward, W.D. 1963. Auditory fatigue and masking. Pp. 240-286 in *Modern Developments in Audiology,* J. Jerger, Ed. Academic Press, New York, New York.

Waring, G.T., R.M. Pace, J.M. Quintal, C.P. Fairfield and K. Maze-Foley, Eds. 2003. North Atlantic right whale (*Eubalaena glacialis*): Western Stock. Pp. 6-13 in *U.S. Atlantic and Gulf of Mexico Marine Mammal Stock Assessments - 2003.* U.S. Department of Commerce, NOAA Technical Memorandum NMFS-NE-182. National Oceanic and Atmospheric Administration, Silver Spring, Maryland.

Wartzok, D., A.N. Popper, J. Gordon, and J. Merrill. 2004. Factors affecting the responses of marine mammals to acoustic disturbance. *Marine Technology Society Journal* 37(4):6-15.

Wasser, S.K., K.C. Hunt, J.L. Brown, K. Cooper, C.M. Crockett, U. Bechert, J.J. Millspaugh, S. Larson, and S.L. Monfort. 2000. A generalized fecal glucocorticoid assay for use in a diverse array of nondomestic mammalian and avian species. *General and Comparative Endocrinology* 120:260-265.

Weiss, J.M. 1968. Effects of coping responses on stress. *Journal of Comparative and Physiological Psychology* 65:251-260.

Weller, D.W., Y.V. Ivaschenko, G.A. Tsidulko, A.M. Burdin and R.L. Brownell, Jr. 2002. *Influence of seismic surveys on Western gray whales off Sakhalin Island, Russia in 2001.* Paper SC/54/BRG14 presented to the IWC Scientific Committee, April 2002. 15 pp. This document is available from the Office of the International Whaling Commission, Cambridge, United Kingdom.

Wells, R.S. 1993. Parental investment patterns of wild bottlenose dolphins. Pp. 58-64 in *Proceedings of the Eighteenth International Marine Animal Trainers' Association Conference*, N.F. Hecker, Ed. International Marine Animal Trainers' Association, Chicago, Illinois.

Wells, R.S. 2003. Dolphin social complexity: Lessons from long-term study and life history. Pp. 32-56 in: *Animal Social Complexity: Intelligence, Culture, and Individualized Societies.* F.B.M. de Waal and P.L. Tyack, Eds. Harvard University Press, Cambridge, Massachusetts.

Whitman, K., A.M. Starfield, H.S. Quadling and C. Packer. 2004. Sustainable trophy hunting of African lions. *Nature* 428:175-178.

Willott, J.F., T. Hnath Chisolm and J.J. Lister. 2001. Modulation of Presbycusis: Current Status and Future Directions. *Audiology & Neuro-Otology* 6(5):231-249.

Wingfield, J.C., C. Breuner, J.D. Jacobs, S. Lynn, D. Maney, M. Ramenofsky and R. Richardson. 1998. Ecological bases of hormone-behavior interactions: the emergency life history stage. *American Zooligist* 38:191-206.

Wingfield, J.C. and M. Ramenofsky. 1999. Hormones and the behavioral ecology of stress. Pp. 1-51 in *Stress Physiology in Animals*, P.H.M. Balm, Ed. Sheffield Academic Press, Sheffield.

Wingfield, J.C. and L.M. Romero. 2001. Adrenocortical responses to stress and their modulation in free-living vertebrates. Pp. 211-234 in *Handbook of Physiology; Section 7: The Endocrine System; Volume IV: Coping with the Environment: Neural and Endocrine Mechanisms*, B.S. McEwen, H.M. Goodman, Eds. Oxford University Press, New York, New York.

York, A.E. 1987. Northern fur seal, *Callorhinus ursinus*, eastern Pacific population (Pribilof Islands, Alaska, and San Miguel Island, California). Pp. 9-21 in *Status, Biology, and Ecology of Fur Seals; Proceedings of an International Symposium and Workshop, Cambridge, England, 23-27 April 1984*, J.P. Croxall and R.L. Gentry, Eds. National Oceanic and Atmospheric Administration (NOAA), National Marine Fisheries Service Techical Report number 51. NOAA, Silver Spring, Maryland.

York, A.E. 1994. The population dynamics of Northern sea lions, 1975-1985. *Marine Mammal Science* 10:38-51.

Appendixes

Appendix A

Committee and Staff Biographies

Douglas Wartzok (Chair) is the vice-provost for academic affairs, dean of the University Graduate School, and professor of biology of Florida International University. Dr. Wartzok served as the associate vice-chancellor for research, dean of the graduate school, and professor of biology at the University of Missouri-St. Louis for 10 years. For the last 30 years, his research has focused on sensory systems of marine mammals and the development of new techniques to study the animals and their use of sensory systems in their natural environment. He and his colleagues have developed acoustic tracking systems for studying seals and radio and satellite tracking systems for studying whales. For 8 years, he edited *Marine Mammal Science*, he is now editor emeritus. Dr. Warzok served on the National Research Council panel that produced the report *Ocean Noise and Marine Mammals* (2003).

Jeanne Altmann is a professor in the Department of Ecology and Evolutionary Biology at Princeton University, a member of the National Academy of Sciences, and a fellow of the Animal Behavior Society. Dr. Altmann pioneered the quantitative study of ecology, demography, and genetics of wild primates and standardized methods for observation of behavior. She carried out groundbreaking work on selection pressures on mothers and established the baseline against which primate life-history studies are compared. Dr. Altmann's current research centers on the magnitude and sources of variability in primate life histories, parental care, and behavioral ontogeny. She is analyzing sources of variability in baboon groups and examining patterns in baboon stability in groups and populations. Her major research interests include nonexperimental research design and analysis and behavioral aspects of conservation. Dr. Altmann has a BA in mathematics and a PhD in behavioral sciences.

Whitlow Au is the chief scientist of the Marine Mammal Research Program at the Hawaii Institute of Marine Biology. He performs research on auditory processes, signal processing, and echolocation primarily in dolphins and whales but also in other species. His research involves psychophysical testing, electrophysiological measurements, underwater acoustics measurements, computer modeling of auditory systems, and artificial neural network computations. Dr. Au is interested in the bioacoustics of marine organisms, from the detection and characterization of sounds to their social and ecologic implications. Dr. Au served as a reviewer on earlier National Research Council reports and is a member of the NRC's Ocean Studies Board.

Katherine Ralls is a senior scientist at the Smithsonian Institution's National Zoological Park. She has broad interests in behavioral ecology, genetics, and conservation of mammals, both terrestrial and marine. Her early research focused on mammalian scent marking, sexual dimorphism, the behavior of captive ungulates, and inbreeding depression in captive mammals and laid the foundations for the genetic and demographic management of captive populations. She is known for her research on endangered and threatened mammals in the western United States, particularly sea otters and kit foxes. Dr. Ralls is a fellow of the Animal Behavior Society and received the Merriam Award from the American Society of Mammalogists and the LaRoe Award from the Society of Conservation Biology. She has served on two previous National Research Council panels.

Anthony M. Starfield is a professor of ecology, evolution and behavior at the University of Minnesota. He obtained a BSc in applied mathematics in 1962 and a PhD in mining engineering in 1965 at the University of the Witwatersrand, Johannesburg. As an applied mathematician, Dr. Starfield uses quantitative modeling as the bridge between science and management, with particular interest in conservation management. His projects have been as diverse as a population model of the Hawaiian monk seal and a model to explore the likely consequences of climate change for the Alaskan tundra. Dr. Starfield has taught workshops on modeling and decision analysis to over 800 conservation scientists and resource managers around the world during the last 10 years. He chaired the annual review committee of the Earth Sciences Division of Lawrence Berkeley Laboratory in 1989 and 1995.

Peter L. Tyack earned his PhD in animal behavior from Rockefeller University in 1982. His research interests include social behavior and vocalizations of cetaceans, including vocal learning and mimicry in their natural communication systems and their responses to human noise. Dr. Tyack has been a senior scientist at the Woods Hole Oceanographic Institution since 1999. He served on National Research Council panels that examined the effects of low-frequency sound on marine mammals in 1994 and 2000.

STAFF

Jennifer Merrill is a senior program officer of the Ocean Studies Board (OSB), and has directed studies since 2001. She earned her PhD in marine and estuarine environmental science from the University of Maryland Center for Environmental Science, Horn Point Laboratory. She directed the National Research Council studies that led to the reports on *Marine Biotechnology in the Twenty-first Century: Problems, Promise, and Products* (2002), *Ocean Noise and Marine Mammals* (2003), and *Exploration of the Seas: Voyage into the Unknown* (2003). In addition, she assisted with the report *Oil in the Sea III* (2003) and the Committee to Review Activities Authorized Under the Methane Hydrate Research and Development Act of 2000, and she serves as the OSB staff contact for the International Council of Scientific Union's Scientific Committee on Oceanic Research.

Sarah Capote is a senior program assistant with the Ocean Studies Board. She earned her BA in history from the University of Wisconsin-Madison in 2001. During her tenure with the board, Ms. Capote has worked on the following reports: *Exploration of the Seas: Voyage into the Unknown* (2003), *Nonnative Oysters in the Chesapeake Bay* (2004), *Future Needs in Deep Submergence Science: Occupied and Unoccupied Vehicles in Basic Ocean Research* (2004), the interim report *Elements of a Science Plan for the North Pacific Research Board* (2004), and *A Vision for the International Polar Year 2007-2008* (2004).

Appendix B

Acronyms

AIM	Acoustic Integration Model
ATOC	Acoustic Thermometry of the Ocean Climate
CEE	Controlled Exposure Experiment
ESA	Endangered Species Act
ESME	Effects of Sound on the Marine Environment
FWS	US Fish and Wildlife Service
GIS	Geographic Information System
IBM	Individual-Based Model
LFA	Low-Frequency Active
MMPA	Marine Mammal Protection Act of 1972
MNPL	Maximum Net Productivity Level
NCEAS	National Center for Ecological Analysis and Synthesis
NEPA	National Environmental Policy Act of 1969
NMFS	National Marine Fisheries Service
NOAA	National Oceanic and Atmospheric Administration

NRC	National Research Council
ONR	Office of Naval Research
OSP	Optimum Sustainable Population
PBR	Potential Biological Removal
PCAD	Population Consequences of Acoustic Disturbance
PTS	Permanent Threshold Shift
SPAWAR	Space and Naval Warfare Systems Center
SURTASS	Surveillance Towed Array Sensor System
TTS	Temporary Threshold Shift

Appendix C

Workshop Agenda and Participants List

Predicting Population Consequences of the Disturbance by Noise on Marine Mammals
National Academy of Sciences
Lecture Hall
2101 Constitution Avenue NW
Washington, DC
March 5-6, 2004

Friday, March 5, 2004

Open Session

Opening remarks, committee introductions, review of workshop format
Douglas Wartzok—Florida International University, Chair
Joanne Bintz—Study Director, Ocean Studies Board

Introduction to Task Statement and Model

PANEL I—INDIVIDUALS TO POPULATIONS
Session Introduction—**Katherine Ralls**
Shripad Tuljapurkar, Dean and Virginia Morrison Professor of Population Studies, Stanford University
Bill Morris, Associate Professor, Department of Biology, Duke University

Bruce Kendall, Assistant Professor, Donald Bren School of Environmental Science & Management, University of California, Santa Barbara

PANEL II—FUNCTIONAL MODULATION OF EFFECTS
Session Introduction—**Jeanne Altmann**
L. Michael Romero, Associate Professor, Department of Biology, Tufts University
Daniel P. Costa, Professor, Department of Ecology and Evolutionary Biology, University of California, Santa Cruz
S.A.L.M. Kooijman, Professor, Department of Theoretical Biology, Vrije Universiteit, Amsterdam

PANEL III—TRANSFER FUNCTION MODELING
Session Introduction—**Anthony Starfield**
Wayne Getz, Department of Environmental Science, Policy and Management, University of California, Berkeley
Gordon Swartzman, Research Professor, Applied Physics Laboratory, University of Washington
Daniel Goodman, Director, Environmental Statistics Group, Montana State University

Saturday, March 6, 2004

Open Session

Opening remarks—**Douglas Wartzok**, Committee Chair

PANEL IV—RESPONSES & MODELS FROM THE MANAGEMENT WORLD
Session Introduction—**Peter Tyack**
Bob Kull, Program Manager, Parsons
Jay Barlow, Program Leader, Southwest Fisheries Science Center, and Adjunct Professor, Scripps Institution of Oceanography
Jean Cochrane, Wildlife Biologist, US Fish and Wildlife Service, Endangered Species Program, Arlington, VA

APPENDIX C

PARTICIPANTS LIST

Committee Members:

Douglas Wartzok (Chair), *Florida International University*
Jeanne Altmann, *Princeton University*
Whitlow Au, *University of Hawaii*
Katherine Ralls, *Smithsonian Institution, National Zoological Park*
Anthony Starfield, *University of Minnesota*
Peter Tyack, *Woods Hole Oceanographic Institution*

Speakers:

Jay Barlow, *Scripps Institution of Oceanography*
Jean Cochrane, *US Fish and Wildlife Service*
Daniel P. Costa, *University of California, Santa Cruz*
Wayne Getz, *University of California, Berkeley*
Daniel Goodman, *Montana State University*
Bruce Kendall, *University of California, Santa Barbara*
S.A.L.M. Kooijman, *Vrije Universiteit, Amsterdam*
Bob Kull, *Parsons*
Bill Morris, *Duke University*
L. Michael Romero, *Tufts University*
Gordon Swartzman, *Applied Physics Laboratory*
Shripad Tuljapurkar, *Stanford University*

Attendees:

Dan Allen, *ChevronTexaco Exploration Production Company*
Laurie Allen, *National Oceanic and Atmospheric Administration*
Charles Bedell, *Murphy Oil Corporation*
Sue Belford, *Jacques Whitford Environment Limited*
Lee Benner, *Minerals Management Service*
Daryl Boness, *Smithsonian Institution, National Zoological Park*
Mel Briscoe, *Office of Naval Research*
Jack Caldwell, *WesternGeco*
Ben Chicoski, *National Oceanic and Atmospheric Administration*
Tara Cox, *Marine Mammal Commission*
Cythia Decker, *Oceanographer of the Navy*

Bridget Ferriss
Phil Fontana, *Veritas Marine Acquisition*
Kellie Foster, *National Oceanic and Atmospheric Administration*
Amy Fraenkel, *Senate Committee on Commerce, Science, and Transportation; Subcommittee on Oceans, Fisheries, and Coast Guard*
Ann Garrett, *National Oceanic and Atmospheric Administration*
Roger Gentry, *National Oceanic and Atmospheric Administration*
Bob Gisiner, *Office of Naval Research*
Mardi Hastings, *Office of Naval Research*
Frank Herr, *Office of Naval Research*
Bob Houtman, *Office of Naval Research*
Mi Ae Kim, *National Marine Fisheries Service*
Karen Kohanowich, *Assistant Secretary of the Navy for Environment*
Anurag Kumar, *Geo-Marine Inc.*
Stan Labak, *Marine Acoustics, Inc.*
David Laist, *Marine Mammal Commission*
Todd McConchie, *George Mason University*
Roger Melton, *ExxonMobil Upstream Research Company*
Harriet Nash, *National Oceanic and Atmospheric Administration*
Patrick O'Brien, *ChevronTexaco Energy Technology Company*
Tim Ragen, *Marine Mammal Commission*
Wallie Rasmunssen, *ExxonMobil Corporation*
Michael Rawson, *Lamont-Doherty Earth Observatory*
Nan Reck, *National Oceanic and Atmospheric Administration*
Naomi Rose, *Humane Society of the United States*
Bill Schmidt, *National Park Service*
Randy Showstack, *Reporter, EOS*
Brandon Southall, *National Oceanic and Atmospheric Administration*
Frank Stone, *Chief of Naval Operations*
Maya Tolstoy, *Lamont-Doherty Earth Observatory*
Kathleen Vigness Raposa, *Marine Acoustics, Inc.*
Erin Vos, *Marine Mammal Commission*
Brian Weitz, *Senate Committee on Commerce, Science, and Transportation; Subcommittee on Oceans, Fisheries, and Coast Guard*
Andrew Wigton, *ExxonMobil Upstream Research Company*
Sheyna Wisdom, *URS Corporation*
Nina Young, *Ocean Conservancy*

Staff:

Susan Roberts, *Acting Board Director*
Joanne Bintz, *Study Director*
Jennifer Merrill, *Study Director*
Sarah Capote, *Program Assistant*

Appendix D

Draft Conceptual Plan for Workshop Discussion

DEFINITION OF THE PROBLEM

Throughout human history oceans have been important for transportation and commerce, biological and physical resource extraction, and defense. However, the vast expanse of the oceans precluded significant human impact until the coming of the industrial revolution. The transition from wind driven to mechanized shipping, was the first step in a continued increase in the initially unintentional and subsequently, with the development of sonar, intentional introduction of sound into the ocean. Because of the low loss characteristics of sound transmission, compared to light transmission, the use of sound had developed evolutionarily as the predominant long-range sensory modality for marine species. Thus as human use of the oceans increased with a concomitant increase in anthropogenic sound in the ocean, the conflict with evolutionarily adapted marine animals sound sensing systems was inevitable.

Over 90 percent of the global trade is transported by sea. Shipping is the dominant sound in the world's oceans at between 5 and 500 Hz. At other frequencies, anthropogenic noise does not predominate in the ocean sound energy budget, but can have important local impacts. For instance, seismic air guns associated with geophysical exploration for locating new oil and gas deposits run hundreds of thousands of miles of survey lines in just the Gulf of Mexico each year. In addition, commercial sonar systems are on all but the smallest pleasure craft. These sonars allow for safer boating and shipping, and more productive fishing. Military sonar systems are important for national defense.

This intentional and unintentional introduction of sound in the ocean associated with activities beneficial to humans must be balanced against known deleterious effects on marine mammals. Strandings of beaked whales in certain environments are clearly associated with the use of mid-range tactical military sonar. There are documented behavioral responses of beluga whales to icebreakers 50 km away. Gray whales and killer whales have shown multi-year abandonment of critical habitats in response to anthropogenic noise. Although there are many documented, clearly discernable responses of marine mammals to anthropogenic sound, reactions are typically subtle, consisting of shorter surfacings, shorter dives, fewer blows per surfacing, longer intervals between blows, ceasing or increasing vocalizations, shortening or lengthening duration of vocalizations, and changing frequency or intensity of vocalizations. Although some of these changes become statistically significant in given exposures, it remains unknown when and how these changes translate into biologically significant effects at either the individual or the population level.

The basic goal of marine mammal conservation is to prevent human activities from threatening marine mammal populations. The threat from commercial whaling was obvious, but it is harder to estimate the population consequences of activities that have less immediately dramatic outcomes, such as those with indirect or small but persistent effects. The life histories and habitat of marine mammals compounds these problems. Marine mammals are long lived and slow to mature. Many species have long periods of dependency. They are highly social and show behavioral plasticity, with complex development of behavior. Furthermore, many of these behaviors occur underwater where they are difficult to document. This makes it particularly difficult to estimate the effects that a short term exposure may have as it ripples through the lifetime of an individual, or as effects on different individuals ripple through the population. Even extreme effects, including death, are not necessarily observed.

The status of any population is the consequence of the accumulation of many effects; resulting in marginal changes in survival and reproduction over time. In addition, the end result is often so far removed in time from the proximate causal events that they cannot simply be traced post hoc. The existence of several comparable populations with different status and different exposure can be used to reduce the number of candidate primary causes of the decline. However, often such comparative populations are lacking.

One way around this conundrum, well tested for issues of human health, is to study how individuals respond to exposure in the short term. Behavior and physiology are rapid response systems evolved to compensate for environmental variation within established limits. A standard method to evaluate risks of exposure to chemicals involves analyzing the short-term physiological responses to specific doses of a compound. Similar studies have been conducted to investigate how marine mammals respond to known exposures to sound. The goal of the NRC Committee on Characterizing Biologically Significant Marine Mammal Behavior is to develop a framework to relate short term acoustic dose:behavioral response relationships to potential population consequences.

HISTORY OF NRC REPORTS

The NRC has produced three reports on the effects of noise on marine mammals, in 1994, 2000 and 2003. The primary goal of the 1994 report was to recommend research on this topic, but the report noted that regulation of marine mammal research impeded critical research, and the report had an entire chapter on regulatory burdens. This chapter of the 1994 report focused especially on harassment of marine mammals. It pointed out that:

> Logically, the term harassment would refer to a human action that causes an adverse effect on the well-being of an individual animal or (potentially) a population of animals. However, "the term 'harass' has been interpreted through practice to include any action that results in an observable change in the behavior of a marine mammal. . . ." (Swartz and Hofman, 1991, p. 27)

> As researchers develop more sophisticated methods for measuring the behavior and physiology of marine mammals in the field (i.e. via telemetry), it is likely that detectable reactions, however minor and brief, will be documented at lower and lower received levels of human-made sound. . . . In that case, subtle and brief reactions are likely to have no effect on the well being of marine mammal individuals or populations. (Swartz and Hofman, 1991, p. 28)

The 2000 NRC report also has a chapter on regulatory issues focusing on acoustic harassment. This chapter continued to emphasize the importance of a criterion for significance of disruption of behavior: "It does not make sense to regulate minor changes in behavior having no adverse impact; rather regulations must focus on significant disruption of behaviors critical to survival and reproduction ..." (Swartz and Hofman, 1991, p 68). It went on to suggest a redefinition of Level B harassment as follows:

> Level B—has the potential to disturb a marine mammal or marine mammal stock in the wild by causing meaningful disruption of biologically significant activities, including, but not limited to, migration, breeding, care of young, predator avoidance or defense, and feeding. (Swartz and Hofman, 1991, p. 69)

The third report of the NRC, *Ocean Noise and Marine Mammals (2003)*, attempted to look at the world ocean noise budget between 1 and 200,000 Hz with particular attention to habitats that were important to marine mammals. The basic question the report tried to address was: What is the overall impact of human-made sound on the marine environment? The somewhat unsatisfactory answer was that the overall impact is unknown, but there is cause for concern. Other than shipping, the overall energy contribution of anthropogenic sound to the ocean noise budget is insignificant. However, total energy contribution is not the best currency to use in determining potential impact of human-made sound on marine organisms. The report made a number of recommendations with the overarching one being the need to better understand the characteristics of ocean noise, particularly from man-made sources and its potential impacts on marine life, especially those that may have population level consequences.

STATEMENT OF TASK

The statement of task for the present NRC Committee, the *Committee on Characterizing Biologically Significant Marine Mammal Behavior*, picks up on two issues noted above: the difference between statistically significant and biologically significant changes in behavior; and linking those short-term behavioral changes to possible population level consequences. The term "biologically significant" enjoys wide use in conservation and management literature, and increasingly in regulatory agency guidelines, but has not been well defined. The committee has been tasked to define "biologically significant" within the context of marine mammal behavioral responses to ocean acoustic sources with particular reference to those responses affecting marine mammal populations. The committee will produce a brief report that reviews and characterizes the current scientific understanding of when animal behavior modifications induced by transient and non-transient ocean acoustic sources, individually or cumulatively, could threaten marine mammal stocks. Recommendations will be based on input from a scientific workshop, consideration of the relevant literature, and other sources.

GOAL, PROPERTIES AND OUTPUT

Develop a conceptual framework and produce a practical process to help regulators assess the risk that specific acoustic sources will have negative impacts on a marine mammal population by disrupting normal behavioral patterns.

Desirable properties of such a process include one that is: accurate; precautionary and becomes more precautionary with greater uncertainty in the potential population level effects of the induced behavioral changes; is simple and transparent to the public, legal staff, and congress; leads to an iterative process which will improve risk estimates as data improve; is able to evaluate cumulative impacts of multiple low level disturbances; and ends up with a small number of parameters that are easy to estimate.

COMMITTEE CONCEPTUAL APPROACH

We propose a process to link acoustic stimuli to behavioral responses to functional outcomes of responses integrated over daily and seasonal cycles in a way that links to life history models. This sequence of stages is essential to link population models, which for seasonal breeders are typically structured on an annual basis, with studies that relate acoustic exposure to behavioral response, that typically work on time scales of hours.

Table D-1 diagrams our approach. On the left we characterize the acoustic features of the sound stimulus of interest. The first stage of our framework involves a transfer function to predict behavioral responses to this sound. Ideally this function derives from controlled exposure experiments, supplemented by observational or correlational studies. This transfer function may vary depending upon the species, season, location, and age-sex of the subject. In the absence of data for the precise situation of interest, marine mammals should be grouped in this stage of the framework by their hearing capabilities, and only data from the same ear type should be used.

TABLE D-1 Transfer functions weighted by season, location, demographic characteristics. Topics highlighted in **bold** were emphasized at the workshop.

Sound	X-Fcn 1	Behavior Change	X-Fcn 2	Function Impacted	**X-Fcn 3**	Population Effect
Transformation Function Modeling						
Freq		Orientation		Life		Survival
Duration		Breathing		Migration		Children
Level		Vocalizing		Feeding		Grandchildren
Source		Diving		Breeding		
Duty Cycle		Resting		Nurturing		
		Mother-Infant		Response to Predator		
		Spatial relationships	**Homeostasis/Risk Factor (Allostasis)**			
		Avoidance	**Time and Energy**			

The output of the first transfer function predicts changes in observable behaviors or physiological measures as a function of sound exposure. The second stage of our framework must evaluate how much these changes in behavior compromise processes that are widely recognized as critical to life history. Where possible, we propose to break down these functional consequences into two time scales—diurnal and seasonal. Most marine mammals respond to diurnal changes with a cycle of activities that suggests the validity of integrating short term functional consequences over a minimum duration typical of the activity in undisturbed animals up to durations of 24 h when possible. These time scales can be studied with behavioral observations or tagging methods. Most marine mammals also show strong seasonal variations in behavior and physiology. As a first cut, our framework will then sum expected daily consequences over each season, depending upon expected exposure schedule to the sound of interest.

The output of the second transfer function defines over a season, the extent to which exposure to a sound may have interfered with the subject's ability to perform behavioral functions that may be critical to survival, growth, and reproduction. The third stage of our framework must estimate what impact this interference may have at the population level. We propose that this stage involves matrix population models structured to stratify each season by the amount of interference. Ideally this would involve models where there is some basis for estimating exposure and thus amount of interference for each individual or age-sex class, depending upon how the model is structured. The function relating interference to population effect ideally would derive from several years of observation of survival and reproduction in a population where effects of exposure can be predicted. For the purposes of this report, we will need to develop a preliminary method to estimate the likelihood of population effect.

SOUND

Ocean acoustic sounds can have a wide range of effects on marine mammals varying from minor annoyances to potentially deleterious effects on a population level. The sources of acoustic noise have been well described in the 2003 National Research Council's (NRC) Ocean Studies Board report, which also described a variety of effects of noise on marine mammals. The discussion of the effects of noise on marine mammals in the 2003 NRC report concentrated on individual marine mammals with the implication that if enough individuals are affected in the same manner, then the population will be affected. In this discussion, the focus will be on the effects of ocean acoustics that will have negative consequence on marine mammals on the population level.

We will attempt to understand how different acoustic sources could modify behavior and hinder marine mammals from performing critical functions that could eventually have an impact on the population level. There are many questions concerning how acoustic signals can modify behavior on a time scale that would affect a population of marine mammals. Among various parameters of acoustic signals that should be considered include bandwidth, frequency range, intensity, modulation type, modulation rate, duration and duty cycles need to be considered. However, at our current level of understanding there is little understanding how any of these parameters, whether individually or corporately can affect or modify marine mammal behavior. Even in a simple case, we would expect that a narrow-band acoustic source will have little effectiveness in disturbing a dolphin's ability to echolocate. Then the question is how broad in bandwidth does the acoustic interference need to be to disrupt or interfere with a dolphin's ability to echolocate? There are many similar questions to which there are no obvious answers.

BEHAVIOR

Behavioral changes typically occur over time ranges of minutes to hours. The responses often increase monotonically with increasing signal intensity, but such changes are rarely linear. They are also strongly influenced by other signal characteristics such as frequency, rise time, duty cycle, novelty, and total energy content. The variability in behavioral responses is as likely due to changes in the state, condition, demographic status, or location of the animal as to characteristics of the sound source. Repeated presentations of the signal typically result in habituation in which the response is not as pronounced to subsequent signal presentations, but the converse can also occur in which the response becomes greater on subsequent presentations of the same signal, a condition known as sensitizitation. Individual variability of animals significantly reduces the capability of predicting behavioral change in response to acoustic stimuli.

FUNCTION

All organisms must perform a set of behavioral and physiological functions in order to survive, grow, and reproduce. Marine mammals must have effective ways to avoid predation, feed, breed, and take care of their young. Many species migrate over long distances, and all must orient on smaller scales. Many pelagic species dive between the surface where they must breathe and great depths where they find and consume prey.

Each of these behavioral activities may be affected by acoustic interference in different ways with different functional consequences. The main costs of interference are risks of injury, opportunity costs due to not detecting a signal, and costs of lost time and extra energy expenditure. If a diving animal responds to sound in a way that pushes the limits of diving physiology, the behavioral response itself could cause injury. If noise stimulates seals to stampede on a beach, or stimulates a cetacean to strand, this could cause death or injury. Similarly if an animal fails to detect an oncoming predator because of interfering noise, it could be killed or injured.

Interpreting the indirect effects where behavioral responses to sound may injure or kill a marine mammal is straightforward. The other costs of lost opportunities, time, or energy require more interpretation to infer the consequences. If an animal incorrectly responds to a noise as if it were a predator, this response entails the costs of lost time and energy. A migrating animal could be affected in two different ways. If it uses acoustic cues to orient for migration, exposure to noise sufficient to mask these cues might interfere with orientation. Some migrating animals avoid exposure to noise; this deflection costs time and energy. If exposure to noise interferes with feeding, the primary costs are time lost if prey items are missed, and energy costs of lost prey intake and potentially increased costs of locomotion. The likely costs of noise to breeding and parental care both involve the costs of not detecting signals and the energy and time costs of any mechanisms they may have for compensating for noise to improve the probability of signal detection in noise. However, the consequences differ. In species that use acoustic communication in the mating system, a female might in the worst case fail to find a mate while she was receptive. This problem is likely to be worst for depleted populations that do not aggregate in mating centers. Noise may also interfere with the process by which males compete during the breeding season, by which females select a mate. All marine mammal young are dependent upon parental care. Many species use acoustic communication both to maintain contact between mother and young, and also for mother-offspring recognition. If increased noise prevented or delayed mother and young from reuniting after a separation, this could have negative consequences for the young. Many marine mammals learn their vocalizations. We are only just beginning to understand the intricacies of vocal development in marine mammals, but increased noise might interfere with development of a fully functional system of vocal communication.

OPERATIONAL PLANS

The Committee held its first meeting 6–8 October 2003 at the National Academy of Sciences in Washington, D.C. and prepared this document conceptualizing and outlining a proposed approach to addressing the statement of task. The committee also identified those areas in which it needed assistance in completing the model leading from stimulus through a determination of biologically significant behavioral change to a population level effect. Four primary areas where additional expertise was needed were identified. For each of those areas, experts will be identified and invited to the next meeting of the committee on 5–8 March 2004, again at the National Academy of Sciences in Washington, D.C. That meeting will begin with a two day workshop. On the first day each of the invited experts will make a 15 minute presentation on how the gaps in the model can be bridged and how the deficiencies in the model can be rectified. On the morning of the second day, the experts within each area will meet together with one member of the committee to put together a synthesis and improvement of the individual presentations of the day before. In the afternoon, each of the four working groups will make a presentation to the full committee. The committee will spend the final two days in closed session writing the report.

TRANSFER FUNCTION WORKING GROUP

The overall purpose of the proposed model is three-fold. The first two purposes derive directly from the statement of task, identifying biologically significant behavioral changes and linking those changes to population level effects. The third purpose is to assist regulators in determining the likelihood that a given stimulus will lead to a specific behavioral change affecting a defined biological function which results in a given change in an identified population parameter. Between each of these operational units there are transfer functions which can be weighted by a variety of external factors such as season, location, and demographic characteristics of the exposed animals. Given the current state of knowledge, the committee recognizes that likelihood factors cannot be categorized on a finer scale than high, moderate or low. The Transfer Function Modeling expert group will help the committee turn this heuristic model into an operational one.

SOLICITATION OF PARTICIPATION

We are all too aware of the questions and uncertainty surrounding our task. On the other hand, decisions affecting the fate of these populations must be made. We face the task given to us not with confidence that we can solve all the problems, but rather in the hope that the framework we develop can help to provide a scientific basis for ranking research and management priorities.

We are soliciting your participation not only in helping to fill in significant areas in which the committee lacks sufficient experience or knowledge, but also your perspective, often from a very different background and experience, as to the overall approach of the committee to the statement of task. This model is being presented very much as a work in progress and we hope you will take this opportunity to help the committee to shape this model, or to convince the committee to abandon this model. Thank you.

Appendix E

Scientific and Common Names

Order Carnivora
    Family Felidae
    Puma concolor coryi   Florida Panther
    Family Mustelidae
    Enhydra lutris   Sea otter
    Family Odobenidae
    Odobenus rosmarus   Walrus
    Family Otariidae
    Zalophus californianus   California sea lion
    Eumetopias jubatus   Steller sea lion
    Family Phocidae
    Mirounga augustirostris   Elephant seal
    Halichoerus grypus   Gray seal
    Phoca vitulina   Harbor seal
    Phoca hispida   Ringed seal
    Leptonychotes weddellii   Weddell seal

Order Cetacea
    Family Balaenidae
    Eubalaena glacialis   North Atlantic right whale
    Family Balaenopteridae
    Megaptera novaeangliae   Humpback whale

Family Delphididae
Tursiops truncatus — Bottlenose dolphin
Pseudorca crassidens — False killer whale
Orcinus orca — Killer whale
Family Eschrichtiidae
Eschrichtius robustus — Western gray whale
Family Kogiidae
Kogia sima — Dwarf sperm whale
Kogia breviceps — Pygmy sperm whale
Family Monodontidae
Delphinapterus leucas — Beluga whale (=white whale)
Monodon monoceros — Narwhal
Family Phocoenidae
Phocoena phocoena — Harbor porpoise
Family Physeteridae
Physeter macrocephalus — Sperm whale
Family Ziphiidae
Berardius bairdii — Baird's beaked whale
Mesoplodon densirostris — Blainville's beaked whale
Ziphius cavirostris — Cuvier's beaked whale

Order Gadiformes
　Family Gadidae
　Theragra chalcogramma — Walleye pollock

Order Rodentia
　Family Muridae
　Peromyscus maniculatus — Wild Deer mouse

Order Squamata
　Family Iguanidae
　Amblyrhynchus cristatus — Marine iguana